PROOF AND CONSEQUENCE

PROOF
AND
CONSEQUENCE

An Introduction to Classical Logic

R.E. Jennings and N.A. Friedrich

broadview press

Library and Archives Canada Cataloguing in Publication

Jennings, R. E. (Raymond Earl)
    Proof and consequence: an introduction to classical logic / R.E. Jennings, N.A. Friedrich.

Includes index.
ISBN 1-55111-547-6

    1. Logic, Symbolic and mathematical—Textbooks. I. Friedrich, N. A. (Nicole Amber), 1964-
II. Title.

BC135.J45 2006   160   C2006-900104-9

Broadview Press is an independent, international publishing house, incorporated in 1985. Broadview believes in shared ownership, both with its employees and with the general public; since the year 2000 Broadview shares have traded publicly on the Toronto Venture Exchange under the symbol BDP.

We welcome comments and suggestions regarding any aspect of our publications — please feel free to contact us at the addresses below or at broadview@broadviewpress.com.

North America
PO Box 1243, Peterborough, Ontario, Canada K9J 7H5
PO Box 1015, 3576 California Road, Orchard Park, NY, USA 14127
Tel: (705) 743-8990; Fax: (705) 743-8353
email: customerservice@broadviewpress.com

UK, Ireland, and continental Europe
NBN International, Estover Road, Plymouth, UK PL6 7PY
Tel: 44 (0) 1752 202300; Fax: 44 (0) 1752 202330
email: enquiries@nbninternational.com

Australia and New Zealand
UNIREPS, University of New South Wales
Sydney, NSW, Australia 2052
Tel: 61 2 9664 0999; Fax: 61 2 9664 5420
email: info.press@unsw.edu.au

www.broadviewpress.com

Broadview Press gratefully acknowledges the financial support of the Government of Canada through the Book Publishing Industry Development Program for our publishing activities.

PRINTED IN CANADA

For Mary and Monet

Contents

Foreword

Any project as prolonged as this is bound, in fruition, to excite in its authors a lively sense of relief – and an even livelier sense of debt to those who have been required to put up with it. Foremost among the weary faithful are Joyce and Tony Friedrich on the one hand and Mary Shaw on the other. In the role of creditors we must also acknowledge all of the young logicianers who with cheerful patience, sometimes abandon, took up versions of **Simon** over a period of fifteen trimesters, made demands, formulated wish-lists, and offered friendly encouragement or (rarely) peremptory advice.

This text and software are a part of a much larger scheme called *The Ara Project* in which we plan support, through learning and course management software, and texts, for three levels of logic instruction, from infant stages through late adolescence: from syllogisms to axiomatic systems and the beginnings of model theory. It is gratifying to have come this far with the project, as we can now look forward to the guidance and support of a very much broader family of **Simon** users, both instructors and students. Needless to say, we hope for continued patience from Joyce, Tony, and Mary.

We should particularly like to express our gratitude to Phil Hanson, who many years ago introduced us to Lemmon's text. We are indebted to Dorian Nicholson and Darko Sarenac for their support in the project, and to Bryson Brown and his students, who, for a term, played Guinea logicianers. We thank our colleague and friend, Norman Swartz, for the use of INDEXX, his award-winning indexing software. We wish also to thank Broadview's team of readers and editors for sage lessons, sound advice, and friendly collaboration.

R.J., N.F.

1

General Introduction

1.1 Where Logic Comes From

Logic began as the maidservant of Philosophy. In the infancy of western philosophical thought, among the Ionians of the early sixth century B.C., critical exchanges between philosophers of different persuasions were virtually non-existent. Philosophers merely presented their own conclusions, that the primary element of the world was water, or that it was air, or that it was the unlimited, with such rudimentary empirical evidence as they could, without critical regard for the claims of rival theorists. The first thoroughly rational attempt at a systematic philosophical account of the nature of things is to be found in the remaining fragment of a poem on the nature of being by Parmenides of Elea (b. 515 B.C.E.). Parmenides was almost certainly inspired by the impressive advances in Number theory and Geometry of his Pythagorean teachers to adapt the methods used to such good effect there to the exploration of more fundamental questions. The puzzlement engendered by Parmenides' *argument* against the possibility of plurality not only inspired the ingenious paradoxes of his pupil, Zeno of Elea (b. 469 B.C.E), and sparked the startling atomistic theories of Leukippus and Democritus. It also engendered that conscious preoccupation with human language which would nurture the academic descendents of Socrates (c.470–399 B.C.E), both those taught at Megara, among them Zeno of Citium, the Stoic, (b. 336 B.C.E.) and those who studied with Plato (430–347 B.C.E.) in Athens, among them Aristotle (384–322 B.C.E.). Most of the branches of logic studied by philosophical logicians today, and in particular the two which constitute the matter of this text, have grown from Megarian and Aristotelian rootstock. Aristotle is usually recognised as the first *symbolic* logician, inasmuch as he introduced the use of upper case Greek letters (A, B, Γ *etc.*) to represent classes of objects. The Stoics, or possibly earlier Megarian philosophers introduced the use of ordinals, the first, the second, *etc.*, (in Greek, the lower case letters with macrons, $\bar{\alpha}$, $\bar{\beta}$, $\bar{\gamma}$ *etc.*) to stand for whole sentences.

The form in which the matter of this text is presented is of more recent origin. Its earliest progenitor is Gottlob Frege (1848-1925) who was the first to propound[1] the fundamental constituents of deductive systems in the form in which they are studied today. The approach of Frege and of Russell and Whitehead later[2] was an axiomatic one. They presented a small number of primitive theorems and two rules by which non-primitive theorems were to be derived. The style of their presentation owed much to the example of traditional presentations of Euclidean Geometry. The recognition that human inferential practices were more naturally presented in the form of a set of rules of inference came out of Alfred Tarski's work in formal semantics and Rudolf Carnap's study of formal syntax[3] G. Gentzen was the first[4] to formulate a system of rules of more or less the form in which they are presented in this text.

1.2 The Nature of the Subject

Throughout almost the whole of its history, the primary interest of logicians has lain in the articulation of what could be called Canons of Correct Reasoning: the formulation of rules by which we could assess the arguments of others, and be guided in our own inferences. The vocabulary of *reasoning, argument, inference*, even *truth* and *falsity*, no doubt has its uses, and competent speakers of the language no doubt use

---

1. In *Begriffsschrift*, Jena, 1879.

2. In Whitehead, A.N. and Bertrand Russell, *Principia Mathematica*, 3 vols. Cambridge, 1910-1913; 2d ed., 1925-1927.

3. *Logische Syntax der Sprache*, Vienna, 1934

4. In 'Untersuchungen über das logische Schliessen', *Mathematische Zeitschrift*, xxxix (1934), pp. 176–210, 405–31.

it to some conversational effect, but it has recently become apparent that although we can use such language in conversation, we do not really know what we are talking about. This is simply the language of the folk-theory of logic; we learn to speak it, but the ability to speak it does not confer understanding of what these items are, nor even that they exist. Like all folk-theoretical vocabulary, there is no reason to suppose that it will survive into more sophisticated theories consciously and deliberately constructed. If developments in logical theory were to cast some light upon the notions of ordinary language, that would doubtless be a benefit, but it would hardly be a criticism of logical theory if in the end it could not. Logical theory should be thought of as providing its own precisely defined objects of study, historically related to those of ordinary speech perhaps, but replacing them. It need not be thought a duty of the logician to give an account of the connection. Formal logic has acquired its own very broad range of interests and applications (in, for example, mathematics, computing, information-processing, and engineering): logicians need not justify their existence by reference to ill-defined historical problems. In fact, it is a cultural feature of the community of logical theorists that they value the introduction of new problems equally with discoveries of new solutions to old ones.

On a related subject, instructors will notice that, unlike many other texts, this one offers no mention of an English exclusive *or.* The connection between the connectives of logic and the so-called 'logical' vocabulary of natural language is discussed at length in Appendix B to this volume. It can, of course, be read and discussed, but its inclusion is intended more to forestall the simple-hearted generalities that mark many presentations of the relationship between formal and natural language. The study of natural language connectives offers a rich and revealing gateway to that more general field of research. But as with logic itself, there are grounds for trusting our understanding of things only to the extent that we have ourselves carefully studied them.

1.3 The Aims of this Book

The logicianer[5] at the beginning of a first course in formal logic, need not be more than a very few courses away from confronting known logical problems for which there are no known solutions, or from his or her own invention of new logical theory. Although present-day logical theory presents a vast subject-matter of study and a wide range of methodologies, an imaginative and well-instructed student can find herself or himself in a position to do original research, albeit within probably some specialized area of logic, after a gratifyingly short period of study. Besides this, logical theory lends itself well to the pleasures of ego-free collaborative investigation.

That being said, the pleasure of doing logic will no doubt for many require cultivation, and the mastery of the subject-matter especially at the earliest stages does require patience and diligence. Most students find themselves pleasurably possessed of a brand new skill after a very few weeks of study.

The repeated reference to pleasure is deliberate. The study of logic should be regarded as a pleasure. The creation of proofs of the sort that a book at this level demands, should be as satisfying as the solution of moderately easy or middlingly difficult intellectual puzzles; proofs given as illustrations should be puzzled through and enjoyed for the techniques they reveal. After the earliest stages, the student should be conscious of questions of mathematical style, particularly as the sort of system here developed, although it greatly simplifies the conception of proof, does not of itself enforce good presentational style at every turn.

This book has two aims: Primarily it aims to provide a formally correct but easily understood first course in the study of formal systems. Secondarily it tries to provide a practical means by which a student can assess his own arguments and those of others. The underlying intention of the former aim is that the formal account should serve, with the

---

5. We here re-introduce this useful and attractive sixteenth-century word, which denoted a student of logic.

minimum of retractions and revisions, as a suitable foundation for later studies in formal logic. The secondary aim is promoted by the provision of graded exercises in which the formal methods are used to represent and assess arguments presented in English.

The debt that this text owes to E.J. Lemmon's text *Beginning Logic*, will be evident throughout most of the following chapters. The style of proof presentation is due ultimately to Patrick Suppes.

A few words about Simon

Almost all of the exercises in this text can be completed using the editors included in **Simon**. Many such exercises are already in **Simon**'s editors' catalogues. As you begin each new chapter, reset the chapter number in **Simon** to augment both the catalogue, and, in the case of the proof editor, the resources that **Simon** provides.

2

Classical Propositional Logic 1

2.1 From Arguments to Proofs

2.1.1 Defining the problem

Historically, logicians were interested in *argument*, the familiar notion, that can be summarized roughly as follows:

> An argument is an ensemble of sentences, of which one is designated as the *conclusion* by some such illative as *therefore, accordingly, so, thus* and so on, and the rest are *premisses,* alternatively a complex sentence in which the premisses are designated by some such illative subordinator as *as, since, because* and so on, and the main clause is the conclusion. (*Illation* is the Latinate word for inference, and the words used to indicate inferential structure or direction are therefore called *illatives.* As you will have noticed from this and the previous sentence, the same vocabulary has non-illative uses as well.)

The central historical interest in arguments was a practical one: how to tell a good argument from a bad one. The notions of *good* and *bad* as applied to arguments remain vague, but we get it approximately right if we say that a good argument is one in which *the premisses provide good reasons for the conclusion.* Since our own interests will be narrower and more precisely defined, we needn't be snotty about what constitutes good reasons in the general case. The arguments

Firefly has just foaled;
therefore, she won't be run in the fifth

and

> The carburettor is flooded;
> so pumping the accelerator won't help

can both be regarded as reasonable arguments in their proper spheres. The argument

> Matilda floats;
> therefore Matilda is guilty

was once thought by some to be a good one, and is now thought by most to have been a bad one. Traditionally, even to say what constitutes a good argument has been recognized as a difficult problem to overcome with any precision. But the problem of how to *tell* a good argument from a bad one was construed as the problem of finding a characteristic by which an argument could be seen to be a good argument or a bad one by an inspection of the words occurring in it and an understanding of them. This necessarily narrows the range of arguments that come within their purview. Logicians have been interested in arguments whose goodness or badness turned upon our understanding of the words having occurrences in them, rather than upon the material facts, of, say, horse-racing or auto-mechanics or demonology. In this regard we can distinguish the foregoing arguments from the argument

> My scarf is blue and green;
> therefore my scarf is coloured.

Within this narrower range of cases the premisses provide sufficiently or insufficiently good reasons for the conclusion because of our understanding of the words having occurrences in the premisses and conclusion: in this case, our understanding of the words *blue, green,* and *coloured* is such that the premiss provides a good reason for the conclusion. We need know nothing about the material facts of my scarf or scarves in general to see that the argument is a good one.

In fact the traditional interests of logicians have been even more narrowly focussed. They have wanted not so much a completely general science of argument, as a science applicable throughout all intellectual pursuits. This requirement has diminished logicians' interest in arguments whose goodness depends upon our understanding of colour vocabulary, as that vocabulary does not have sufficiently general application. Logicians have been interested in arguments whose goodness depends upon occurrences of so-called 'logical' words such as *if, then, not, or, and* and so on, such arguments as the following:

> Harry is happy and Gladys is glad;
> therefore, Harry is happy

the goodness of which is evident from our understanding of *and*,

> If Harry is happy, then Sally is sad;
> Harry is happy;
> therefore Sally is sad

the goodness of which we recognize through our understanding of *if . . . then*,

> If Gladys is glad, then her *tchum* is glum;
> her *tchum* is not glum;
> therefore, Gladys is not glad

the goodness of which is apparent from our understanding of *if . . . then* and *not*, and

> *Either* you will party *or* you will pray;
> *if* you pray, *then* you will offend your friend;
> *if* you party, *then* you will anger your statue;
> you will *not* anger your statue;
> therefore, you will offend your friend,

which, with a little attention, we can see to be a good argument solely because of our common understanding of the words italicised in it, not because of our understanding of any of the other vocabulary present. To consider only the first, we can readily see that any argument would be a good argument that has as a premiss two (declarative) sentences joined with *and* and that has one of those sentences as a conclusion. Using A and B to take the places of sentences, we can see that

> A and B;
> therefore A

will be a good argument whatever sentences "A" and "B" are. Similarly,

If A then B;
A;
therefore B

and

If A then B;
not B;
therefore not A

in which 'not B' ('not A') is understood as the sentence in the B (A) place with a *not* properly inserted. On the same terms, we can see, with a little reflection that any argument

Either A or B;
if A then C;
If B then D;
not D;
therefore C.

would be a good argument.

Now the theory that we are about to explore has vocabulary that can harmlessly be regarded as mathematical counterparts of the English 'logical' vocabulary of those examples, but we should note at the outset that the choice of vocabulary, though not arbitrary, is nevertheless artificially restricted. There is no reason why, for example, *unless*, or *only if* should not been mentioned in the list even for the part of logical theory to be examined. And other pieces of English vocabulary such as *necessarily, knows, believes, ought, eventually, since* have also been elected as 'logical' subject matter, and mathematical counterparts for them introduced and studied. Even within this text, we will eventually expand our selection to include such words as *all, some, same* and *the*. Against the background of such a selection of logical vocabulary, we can speak of an argument's being *logically good*, the logicality of its goodness being relative to the vocabulary selected.

2.1.2 Our problem stated

The problem that confronts us can be stated in terms that are independent of the question as to which vocabulary we have actually chosen to regard as logical vocabulary. Suppose that we are given an argument,say, sealed in an envelope so that we do not know whether it is a good one or a bad one, only that it is one or the other. We would like to have procedures to hand that would ensure that after finitely many steps (the first of which will presumably be to open the envelope), we will be able to answer the question: is it a good argument or is it a bad one according to our understanding of its logical vocabulary? Such a procedure, one by which we know after finitely many steps whether an object (in this case an argument) is a member or is not a member of a given set (in this case, the set of good arguments), is said to be an *effective* procedure.

We can restate the original objective as two objectives: a positive one and a negative one. If an argument is a good one we would like a finite procedure that will tell us that it is a good one; if the argument is a bad one, we would like a finite procedure that will tell us that it is bad. In this chapter we consider a procedure by which if an argument is a logically good one it can be shown to be good; we leave the negative objective for later consideration.

2.1.3 The intuitive idea of a proof

Intuitively, to say that premisses provide logically good reasons for a conclusion can be thought of as the claim that the information of the conclusion is implicitly present in the premisses. (The information that my scarf is coloured is a part of the information that my scarf is green and blue.) The positive procedure can be thought of intuitively as a procedure for extracting the information of the conclusion from the information of the premisses. Since we haven't defined the notion of *information*, that must remain a purely intuitive understanding of what we are about. Officially our procedure consists simply in this: we provide a finite system of intuitively correct rules that let us write down the premisses, and then let us write down later sentences of a certain sort given that we have already written down sentences of a certain sort. I say 'officially', but in fact our procedure does not 'officially' mention inscription at all. It merely licenses or permits certain sequences of sentences. There is a rule that permits any premiss to be a member of a sequence, and rules that permit certain sequences of sentences to be extended to sequences containing certain other sentences. If the rules permit the extension of a sequence of sentences containing the premisses of an argument to a sequence of sentences of a certain sort containing the conclusion, then the rules will be said to license the argument. The resulting sequence of sentences will be said to be a proof (for the system of rules) of the conclusion of the argument from its premisses. The rules themselves will correspond to kinds of inference that we would regard as intuitively acceptable for the natural-language counterparts of the logical vocabulary we adopt. So, to revert to the language of inscriptions, our procedure will be this: in accordance with one of the rules, we will write down the premisses; then we will try to show that the rules eventually let us write down the conclusion. But before we can consider the rules, or make the idea of a proof more precise, we must provide the language to which the rules will be applied.

2.1.4 The language of the system

English connective vocabulary mostly comes in pairs:

if . . . then – – –
either . . . or – – – –
both . . . and – – –.

We can call the first member of each pair its *prefix*, and the second its *infix*. Our system will adopt symbols only for the infix position: we will understand 'A → B' ('A arrow B') as the counterpart of *if A, then B*. We will understand 'A ∨ B' ('A vee B')[1] as the counterpart of *either A or B*. We will understand 'A ∧ B' ('A cap B') as the counterpart of *Both A and B*. A sentence 'A → B' is called a *conditional*, 'A' its *antecedent*, and 'B' its *consequent*. A sentence 'A ∨ B' is called a *disjunction*, and 'A' and 'B' its *disjuncts*. A sentence 'A ∧ B' is called a *conjunction*, and 'A' and 'B' its *conjuncts*. Each of →, ∧, and ∨ are called *connectives*, and because they join two sentences, are said to be *binary connectives*.

---

1. The ∨ was initially introduced to suggest the word *vel*, one of numerous *or* words of Latin.

In English, negations are variously expressed, but not usually in such a way that we can conveniently preserve some vestige of the English form in our logical language. We will understand '¬A' as the *negation* of 'A' and read it 'not-A'. Perhaps the closest we come to this in English is in the stilted academese of 'It is not the case that . . .'

Notice that in describing our logical language, we have been using the English language as the vehicle of the description. Or rather we have been using an augmentation of English, for as we earlier used 'A', 'B', 'C' and 'D' as symbols standing for anonymous sentences of the English language, so we continue to use them to stand for anonymous sentences of our logical language. In this role they are called *metalogical variables*. In such an exercise it is usual to refer to the language being described as *the object language* and the language in which the description is given as *the metalanguage*. If we describe English in French, then English is the object language and French the metalanguage of the description. If we describe English in English, then English is both the object language and the metalanguage of the description. We will give a more precise account of our object language later; here it will be sufficient to say that it has an indefinitely large set of unanalysable sentences called *atoms*, which for all official purposes will be the symbols

$P_1, P_2, \ldots P_3 \ldots$

but which for day-to-day work can be written as

$P, Q, R, \ldots$

We will also permit ourselves to make free use of parentheses '(' and ')' on much the same basis as we would use them in algebraic constructions. We will define their role more rigorously later.

Now with just the range of symbols so far introduced we can construct in our language sentences of any finite complexity, so, for example, sentences requiring any finite number of nested parentheses. All of the following are sentences of the language:

P

P ∨ Q

(P ∨ Q) → (R → (Q ∨ S))

¬((¬¬¬¬¬P ∨ ¬¬Q) → ¬¬¬¬(¬(R → (¬¬¬¬Q ∨ ¬¬S))) → ¬(¬(S ∨ ¬¬¬Q) → ¬R))).

We can also represent the argument that we schematized earlier as

> Either A or B;
> if A then C;
> If B then D;
> not D;
> therefore C.

directly in our language as

> P ∨ Q
> P → R
> Q → S
> ¬S
> therefore R.

In fact we can go a little further: the metalogical symbol '⊢$_L$' (read 'turnstile sub L') represents the relationship of provability for a system L of rules. The expression

(Premises)   (conclusion

$$P \lor Q, P \rightarrow R, Q \rightarrow S, \neg S \vdash_L R \quad = \text{ sequent}$$

is to be understood as the metalinguistic claim that there is an L-proof (that is, a proof using the rules of the system L) of the sentence on the right of the turnstile (R) from the ensemble of sentences on the left of it (P ∨ Q, P → R, Q → S, ¬S). Such an expression is called a *sequent*. The sentences listed on its left are its *premises*; the sentence on its right is its *conclusion*. We demonstrate the truth of the claim of the sequent by constructing a proof of its conclusion from its premises using the rules of the system L. We demonstrate its falsity by demonstrating that no such proof exists. And here we should add an historical comment that is also an important statement about what follows.

> The study of the relationship symbolized by the turnstile is the whole subject matter of this text, and the central subject matter of logical theory. Bear in mind that the relation symbolized by $\vdash_L$ is not restricted to finite sets, and can be studied in its own right without this restriction. However, the expressions that we call *sequents* claim provability of a conclusion from a finite set of premisses.

We are not here learning to reason or to argue well. We are studying the relationship of L-provability. To be sure, we must learn how to construct L-proofs, but our aim is also to study the properties of L-provability more generally. To this end, we will sometimes use upper-case Greek letters, Γ (gamma), Δ (delta), Σ (sigma), and so on, for arbitrary ensembles of sentences, and write

$$\Sigma \vdash_L A$$

to be read 'there is an L-proof of the sentence A from the ensemble, sigma of sentences.' We will sometimes have occasion to write '∅' for the empty or null ensemble.

2.1.5 The documentation of proofs

An L-proof, though it must satisfy other conditions, is just a finite sequence of sentences of the language of L. Nevertheless, in the presentation of a proof, it is usual to provide additional documentation: to number the entries, and to justify entries by the mention of the rules and the previous entries that have brought those rules into use. These all help the reader follow the proof, but also serve as reminders to its author. In the system here to be presented, an additional item of documentation is included. At the extreme left of every line, we keep track (by their line number) of the assumptions from which the entry at that line has been proved. At the last line of the proof, the entry, which in this case is the sentence to the right of the $\vdash$, must have been proved from the sentences on the left of the $\vdash$. Therefore, at the last line of the proof, all of the assumptions listed must be premisses, that is, sentences occurring to the left of the $\vdash$. In the course of a proof, the sentence being proved may occur numerously as an entry, but only when no assumptions are listed on the left except premisses can the proof end. We should note that not every premiss need be listed, but no non-premiss can be listed at the last line of a proof.

2.2 The Rules of L

Since the system to be presented is due to E. J. Lemmon, it is fitting that we should refer to it as the system L. However, since it is the only system under consideration, we dispense with the subscript throughout.

2.2.1 The Rule of Assumption (A)

Later we will give ourselves the means of abbreviating proofs; nevertheless, every unabbreviated proof, and so we may say, *officially* every proof, begins with an application of the *Rule of Assumption*. This rule permits us to write down any sentence of our language as a line of a proof. In a proof of the sequent given as an example, $P \vee Q, P \rightarrow R, Q \rightarrow S, \neg S \vdash R$, the first four lines would be the following

Assumption line. →

1	(1)	$P \vee Q$	A
2	(2)	$P \rightarrow R$	A
3	(3)	$Q \rightarrow S$	A
4	(4)	$\neg S$	A

Notice that because the use of the Rule of Assumption does not require any previous entries in the proof, *no line numbers are mentioned in the justification.* Notice also that when the Rule of Assumption is invoked to justify an entry in a proof, *the line number listed at the left is the line number of the entry itself.* Recall that in general, that entry indicates the assumption from which the entry at the line has at this stage actually been proved. This fact invites one or two comments about the Rule of Assumption, about proofs and about provability in general.

2.2.2 Some observations

Because the entries at the extreme left of a line tell us from which assumptions the entry at that line has been proved, we can think of each line as having a *corresponding sequent*. The sequent corresponding to a line of a proof is the sequent having the listed assumptions of the line to the left of its turnstile (that is, as its premisses), and the entry of the line to the right, (that is, as its conclusion).

In speaking of the relationship of provability, it is usual to distinguish two kinds of properties, those that are consequences of the particulars of the language (its having ¬, ∨, ∧, →, and so on rather than some other connectives), and those that are entirely independent of such particulars. The latter are usually called *structural properties*. Now the Rule of Assumption represents one such structural property of classical provability, namely the property of *reflexivity*. In its simplest statement as a property of ⊢, reflexivity is the property that every sentence is provable from itself:

> For every sentence A, A ⊢ A.

To say that L is a *classical* system is to say *inter alia* that reflexivity is a property of ⊢$_L$. Thus every application of the Rule of Assumption is a proof of a sentence from itself. Finally, recalling the earlier remark that the last line of a proof must list only, but need not list all premisses of the sequent proved, we can give the property of reflexivity its usual more general statement:

> For every ensemble Σ, and every sentence A, if A is an element of Σ, then Σ ⊢ A.

2.2.3 Modus Ponendo Ponens (MPP)

Pono, ponere is the Latin verb meaning *to place* or *to lay down*. Like our verb *to lay down*, it also had a use meaning *to assert*. The word *modus* means *way* or *mode*. Thus the tag *modus ponendo ponens* referred to a mode of inference by which, given a conditional, one asserted the consequent of a conditional by asserting its antecedent. The tag is a fitting name for the first language-particular rule of the system L.

A proof that includes a proof of A → B from the ensemble Σ, and a proof of A from the ensemble Γ, can be extended to a proof of B from the ensemble Σ,Γ.

In its simplest application, its use in a proof would look like the following proof of the sequent

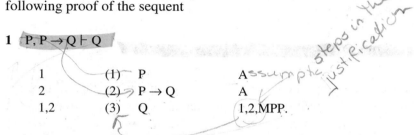

1 P, P → Q ⊢ Q

1	(1)	P	Assumption
2	(2)	P → Q	A
1,2	(3)	Q	1,2,MPP.

With MPP as with other rules, we refer to the entries cited (by line number) in the justification as *the input to the rule* and the entry justified as *the output of the rule*. As with all rules that take individual entries as input, it does not matter in which order the entries occur. The following would be an equally correct application of the rule

1	(1)	P → Q	A
2	(2)	P	A
1,2	(3)	Q	1,2,MPP.

Even with just the Rule of Assumption and MPP, we can construct less simple proofs, as for example, a proof of the sequents

2 P → Q, Q → R, P ⊢ R

1	(1)	P → Q	A
2	(2)	Q → R	A
3	(3)	P	A
1,3	(4)	Q	1,3,MPP
1,2,3	(6)	R	2,4,MPP

and

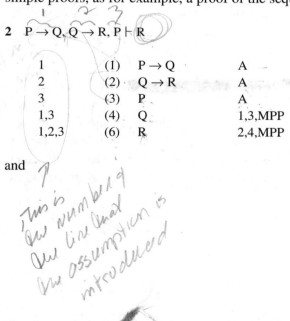

3 $P \rightarrow (Q \rightarrow R), P, Q \vdash R$

1	(1)	$P \rightarrow (Q \rightarrow R)$	A
2	(2)	P	A
3	(3)	Q	A
1,2	(4)	$Q \rightarrow R$	1,2,MPP
1,2,3	(5)	R	3,4,MPP.

2.2.4 The Rule of Double Negation (DN)

It is not the case that the tap is not dripping only if the tap is dripping. This property of natural language negation that a second occurrence, as it were, undoes the work of a first is shared by the negation of our logical language. The Rule of Double Negation is the following:

> A proof of a sentence $\neg\neg A$ from an ensemble Σ can be extended to a proof of A from the ensemble Σ.

The simplest application of DN is the proof of the sequent

4 $\neg\neg P \vdash P$

1	(1)	$\neg\neg P$	A
1	(2)	P	1,DN.

Note that, unlike MPP, DN has only one input entry. Note also that the negation of a sentence A is the sentence A with a negation added. Thus the negation of $\neg P$ is $\neg\neg P$.

2.2.5 The Rule of Conditional Proof (CP)

This rule is intended to capture part of the intuitive connection between logical instances of the conditional $A \rightarrow B$, and the provability of B from A. Part of this connection is already captured in the rule MPP which yields a proof of B from proofs of $A \rightarrow B$ and A. CP gives us the converse of that connection:

A proof of B from an ensemble Σ,A can be extended to a proof of A → B from Σ.

Practically, the rule gives us the means of proving a conditional from a set of assumptions by the following method: first assume the antecedent (by the Rule of Assumption); second prove the consequent from the now augmented set of assumptions. Unlike any of the rules we have introduced so far, *CP takes a proof as its input.* The justification requires a reference to the proof. We mention (a) the line where the proof begins, that is, where the antecedent is assumed, and (b) the line where the consequent later appears. Note that *any such proof must begin with the assumption of the antecedent.* Do not be tempted to cite, as the beginning of a CP proof, a line in which the antecedent is the output of any rule other than the Rule of Assumption.

CP is not the only rule that takes a proof as its input, and therefore requires us to make an assumption. You may find it useful, therefore, to label assumptions parenthetically as in the last proof, to keep track of the intended roles of assumptions. If you are using the **Simon** proof editor to construct the proof, there is an additional benefit to be had from labelling assumptions. This is that the proof checker residing in that **Simon** editor will check whether the assumption is a suitable one for the purpose given in the label. Bear in mind, however, that such a notation does not preclude the later use of the same assumption for some other rule. Notice, in the following proof, that in accordance with the statement of the rule CP, 3 is not listed among the assumptions from which the entry at line (6) has been proved. The assumption of the antecedent is said, at line (6), to have been *discharged.*

5 $P \to (Q \to R) \vdash Q \to (P \to R)$

1	(1)	$P \to (Q \to R)$	A
2	(2)	Q	A (CP)
3	(3)	P	A (CP)
1,3	(4)	$Q \to R$	1,3,MPP
1,2,3	(5)	R	2,4,MPP
1,2	(6)	$P \to R$	3,5,CP
1	(7)	$Q \to (P \to R)$	2,6,CP

Rule of Thumb: In devising a strategy for a proof, you will naturally consider what you are trying prove and what you are trying to prove it from. *If what you are trying to prove is a conditional, then consider using conditional proof.* Study the previously proved sequent in this light. We wanted to prove the conditional $Q \rightarrow (P \rightarrow R)$. So we assumed its antecedent at line (2) for CP. CP requires us then to prove the consequent. But the consequent is itself a conditional. So we elected to prove it by CP and to that end assumed the antecedent for conditional proof at line (3). We proved the consequent of that conditional at line (5); so CP let us write the conditional $P \rightarrow R$ at line (6). But that is the consequent of the first conditional, $Q \rightarrow (P \rightarrow R)$; so CP permitted us to write that conditional at line (7).

Some notes about Simon

Simon already contains almost all of the exercises in the text. Each problem in the following exercise will be found as a partially completed exercise of Chapter 2. In each case just complete the proof.

When using the proof editor of **Simon**, you will first double-click on the workspace provided, and fill in all of the fields for the current line of the proof. When you submit each line, **Simon** will perform two tasks: first, it will check that the syntax of your entries is correct (that its parts are in the right order and properly punctuated); second, it will check that the line is a correct application of the cited rule. If all is well, it will enter the line and mark it 'ok'. If you have made an error, **Simon** will blush in the first field in which errors occur. The 'help' facility will explain your error from **Simon**'s point of view. You have the option of correcting the error or telling **Simon** to accept the line as it is. If you choose this latter option, **Simon** will add the line to the proof, but marked with an 'x'. Be aware that **Simon** is unable to judge the correctness or incorrectness of any later lines that make reference to a line containing an error.

Simon will also recognize a submitted line that completes a correct proof, and adds that line to the proof with a following 'QED' (*Quod erat demonstrandum*: that which was being proved). When **Simon** gives you a 'QED', your proof is complete.

Exercise 2.1

Recall that a proof is a sequence of sentences satisfying certain conditions. The satisfaction of those conditions is exhibited by additional documentation. In the following examples, proofs are provided for sequents, but the documentation has been omitted. In each proof, use **Simon**'s proof editor to fill in the missing documentation.

(a) $P \rightarrow (Q \rightarrow R) \vdash (P \rightarrow Q) \rightarrow (P \rightarrow R)$

$P \rightarrow (Q \rightarrow R)$
$P \rightarrow Q$ Assume any antecedent
P
$Q \rightarrow R$
Q
R
$P \rightarrow R$
$(P \rightarrow Q) \rightarrow (P \rightarrow R)$

(b) $P \rightarrow Q \vdash (Q \rightarrow R) \rightarrow (P \rightarrow R)$

$P \rightarrow Q$
$Q \rightarrow R$
P
Q
R
$P \rightarrow R$
$(Q \rightarrow R) \rightarrow (P \rightarrow R)$.

Exercise 2.2

Use **Simon**'s proof editor to construct a proof for each of the following sequents citing no rules other than A, MPP, DN, and CP.

(a) $P \rightarrow (P \rightarrow Q), P \vdash Q$
(b) $P \rightarrow \neg\neg Q, P \vdash Q$
(c) $P \rightarrow Q, Q \rightarrow R \vdash P \rightarrow R$
(d) $P \rightarrow (Q \rightarrow (R \rightarrow S)) \vdash R \rightarrow (Q \rightarrow (P \rightarrow S))$
(e) $P \vdash (P \rightarrow Q) \rightarrow Q$

2.2.6 The character of the rules so far

Commonly, language-specific rules in a natural-deductive system are two types:

(a)   those that permit an entry with occurrences of a connective from inputs in which the connective does not occur. These are called *Introduction* rules.

(b)   those that permit an entry without occurrences of a connective from inputs in some of which the connective occurs. These are called *Elimination* rules.

Here we have given standard historical names to rules for the →, and for ¬ that could have worn 'introduction' and 'elimination' labels. MPP is an →-elimination rule; CP is an →-introduction rule. DN is a ¬-elimination rule that permits ¬ to be eliminated in pairs. In the cases of the remaining logical vocabulary in our selection, ∧ and ∨, our labelling coincides explicitly with the standard pattern.

2.2.7 The Rule of ∧-introduction (∧I)

The earliest theorists of propositional logic, the Stoics, seem not to have thought *and* (Greek *kai*) to be a logical word at all in any interesting sense of 'logical'. For them there was no inferential difference between the single sentence *Dion is walking and Dion is talking* and the same pair of sentences in tandem. The rules of ∧-introduction and ∧-elimination reflect much the same attitude. The former is:

A proof that includes both a proof of A from ensemble Γ, and a proof of B from ensemble Σ can be extended to a proof of A ∧ B from the ensemble Γ,Σ.

In its simplest application, ∧I permits a proof of

6 P, Q ⊢ P ∧ Q

1	(1)	P	A
2	(2)	Q	A
1,2	(3)	P ∧ Q	1,2,∧I.

As a second example,

7 (P ∧ Q) → R ⊢ P → (Q → R) (Exportation)

1	(1)	(P ∧ Q) → R	A
2	(2)	P	A (CP)
3	(3)	Q	A (CP)
2,3	(4)	P ∧ Q	2,3,∧I
1,2,3	(5)	R	1,4,MPP
1,2	(6)	Q → R	3,5,CP
1	(7)	P → (Q → R)	2,6,CP

2.2.8 The Rule of ∧-elimination (∧E)

The rule ∧E permits the detachment of either conjunct from a conjunction already proved:

A proof of A ∧ B from ensemble Γ can be extended to a proof of A (B) from Γ.

Thus, both of the following proofs are permitted,

8 (P ∧ Q) ⊢ P

1	(1)	P ∧ Q	A
1	(2)	P	1,∧E

9 (P ∧ Q) ⊢ Q

1	(1)	P ∧ Q	A
1	(2)	Q	1,∧E

as is

10 $P \rightarrow (Q \rightarrow R) \vdash (P \wedge Q) \rightarrow R$ (Importation)

1	(1)	$P \rightarrow (Q \rightarrow R)$	A
2	(2)	$P \wedge Q$	A (CP)
2	(3)	P	2,$\wedge$E
1,2	(4)	$Q \rightarrow R$	1,3,MPP
2	(5)	Q	2,$\wedge$E
1,2	(6)	R	4,5,MPP
1	(7)	$(P \wedge Q) \rightarrow R$	2,6,CP.

Taken together, $\wedge$I and $\wedge$E permit proofs of other standard sequents.

11 $P \wedge Q \vdash Q \wedge P$ ($\wedge$-Commutation)

1	(1)	$P \wedge Q$	A
1	(2)	P	1,$\wedge$E
1	(3)	Q	1,$\wedge$E
1	(4)	$Q \wedge P$	2,3,$\wedge$I

12 $Q \rightarrow R \vdash (P \wedge Q) \rightarrow (P \wedge R)$

1	(1)	$Q \rightarrow R$	A
2	(2)	$P \wedge Q$	A (CP)
2	(3)	Q	2,$\wedge$E
1,2	(4)	R	1,3,MPP
2	(5)	P	2,$\wedge$E
1,2	(6)	$P \wedge R$	4,5,$\wedge$I
1	(7)	$(P \wedge Q) \rightarrow (P \wedge R)$	2,6,CP

2.2.9 Monotonicity of provability

We have already touched upon this topic (p. 18) in speaking of the structural property of reflexivity. A provability relation is *monotonic* if and only if it satisfies the condition:

if a sentence A is provable from an ensemble Γ, then A is provable from any ensemble Γ' that includes Γ.

Monotonicity is a structural property of the classical $\vdash$. In the light of the rules, $\wedge$I and $\wedge$E, we can illustrate the property as it affects applications of the rule CP. Consider the following proof of

13 R, R $\rightarrow$ Q $\vdash$ P $\rightarrow$ Q

1	(1)	R	A
2	(2)	R $\rightarrow$ Q	A
3	(3)	P	A (CP)
1,2	(4)	Q	1,2,MPP
1,2	(5)	P $\rightarrow$ Q	3,4,CP

The proof may seem intuitively suspect, since the proof of Q does not seem to require the assumption of P. But notice that what that fact amounts to in this case is just that line (3) is not listed among the entries from which Q has been proved at line 4. In classical logic, the vague idea that the consequent of a conditional must somehow be connected to the antecedent has no place. In truth, the notion has no foundation in the natural language of conditionals either. The construction

I'll be home; so if you come round, I'll be home

is not odd as a piece of English: unconditioned assertions license conditioned ones. The reason why the speaker will be home may have nothing to do with the anticipated visit. In the example, the only deficiency that seems unintuitive, namely the absence of '3' among the list of assumptions at line (4) can be made up in a way that goes nowhere toward making the $\rightarrow$ of our system answer to the vague notion of connectedness of antecedent and consequent. Simply amend the proof making use of $\wedge$I and $\wedge$E.

1	(1)	R	A
2	(2)	R $\rightarrow$ Q	A
3	(3)	P	A (CP)
1,2	(4)	Q	1,2,MPP
1,2,3	(4a)	P $\wedge$ Q	3,4,$\wedge$I
1,2,3	(4b)	Q	4a,$\wedge$E
1,2	(5)	P $\rightarrow$ Q	3,4,CP

Since this procedure consists in a joining and a separating for the sole purpose of creating an official if wholly illusory dependence, it is sometimes referred to as a Reno marriage and divorce. In fact, however, classical logic need not exact this sort of alimony. *We need never use all of the premisses of a sequent in the course of a proof of the sequent.* It is, however, a useful observation about the system L that any proof of a sequent that does not use all of the premisses of the sequent can be expanded to one that does. One need only conjoin all of the premisses by ∧I, and then detach them again by ∧E to have proved each of them individually from all of them taken together.

2.2.10 The Rule of ∨-Introduction (∨I)

∨I requires little discussion. It is

> A proof of A from ensemble Γ can be extended to a proof of A ∨ B (B ∨ A) from the ensemble Γ.

The simplest applications:

14   P ⊢ P ∨ Q

1	(1)	P	A
1	(2)	P ∨ Q	1,∨I

15   P ⊢ Q ∨ P

1	(1)	P	A
1	(2)	Q ∨ P	1,∨I.

A further example:

16   P, (P ∨ Q) → R ⊢ R

1	(1)	P	A
2	(2)	(P ∨ Q) → R	A
1	(3)	P ∨ Q	1,∨I
1,2	(4)	R	2,3,MPP.

2.2.11 The Rule of ∨-Elimination (∨E)

Imagine having the information about a friend that either she is in Paris or she is in Toulouse. What inferences can reliably be made from this information and why? Most obviously we can infer that she is in France, and the reason is that both cities are in that country. That is to say, we can infer that she is in France, because we could infer that from her being in Paris *and* we could infer that from her being in Toulouse. In fact we can infer no more from the disjunction than what we can infer from its disjuncts separately, so the fit is exact. We can infer a sentence from a disjunction if we can infer it from each of the disjuncts separately. The rule of ∨-elimination reflects that inferential fact.

> A proof that includes (a) a proof of a disjunction A ∨ B from ensemble Γ, (b) a proof of C from the ensemble Δ,A and (c) a proof of C from ensemble Σ,B can be extended to a proof of C from Γ,Δ,Σ.

In practical terms, the rule comes into play when an entry in a proof is a disjunction, A ∨ B. To apply the rule to prove some sentence C, first write down A justified by the rule of assumption, and contrive a proof of C from that line together with any previous lines; then write down B justified by the rule of assumption, and contrive a proof of C from that line and any previous line. Then write down C a third time, but now justified by ∨E. In the justification, five lines will be mentioned: the line where A ∨ B occurs, then the lines where each of the two proofs begins and ends. An example will illustrate:

17 P ∨ Q ⊢ Q ∨ P (∨-Commutation)

1	(1)	P ∨ Q	A
2	(2)	P	A (∨E)
2	(3)	Q ∨ P	2,∨I (end of first proof)
4	(4)	Q	A (∨E)
4	(5)	Q ∨ P	4,∨I (end of second proof)
1	(6)	Q ∨ P	1,2,3,4,5,∨E

(handwritten annotations: "-C", "Q∨P must be deduced from Both P and")

We wanted a proof of Q ∨ P from P ∨ Q; so, first we proved Q ∨ P from P; then we proved Q ∨ P from Q. Since it is provable from each of the disjuncts separately, it is provable from P ∨ Q.

Notice the lines cited in the justification of line 6. 1 is the line where the disjunction is entered; 2 and 3 represent the first and last lines respectively of the proof of Q ∨ P from P; 4 and 5 represent the first and last lines respectively of the proof of Q ∨ P from Q.

Now notice the single line mentioned at the extreme left of line 6, that is, 1. This can be arrived at by consulting the lines (1,2,3,4,5) cited in the justification, following this procedure:
Write down all of the assumptions from which 1 was proved. Then write down all of the assumptions from which Q ∨ P was proved at line 3, except the assumption at 2. Then write down all of the assumptions from which Q ∨ P was proved at line 5, except for the assumption at line 4. This procedure leaves us with exactly 1.

Consider a slightly less trivial example also labelled:

18 Q → R ⊢ (P ∨ Q) → (P ∨ R)

1	(1)	Q → R	A
2	(2)	P ∨ Q	A (CP)
3	(3)	P	A (1st for ∨E)
3	(4)	P ∨ R	3,∨I(ends 1st)
5	(5)	Q	A (2nd for ∨E)
1,5	(6)	R	1,5,MPP
1,5	(7)	P ∨ R	6,∨I(ends 2nd)
1,2	(8)	P ∨ R	2,3,4,5,7,∨E
1	(9)	(P ∨ Q) → (P ∨ R)	1,8,CP

and another example, this one with one ∨E proof inside another:

19 P ∨ (Q ∨ R) ⊢ Q ∨ (P ∨ R)

1	(1)	P ∨ (Q ∨ R)	A
2	(2)	P	A (1st outer ∨E)
2	(3)	P ∨ R	2,∨I

(handwritten margin note: "The assumption gets the line number of our line it was introduced on")

2	(4)	$Q \vee (P \vee R)$	3,$\vee$I (ends 1st outer proof)
5	(5)	$Q \vee R$	A (2nd outer $\vee$E)
6	(6)	Q	A (1st inner $\vee$E)
6	(7)	$Q \vee (P \vee R)$	6,$\vee$I (ends 1st inner proof)
8	(8)	R	A (2nd inner $\vee$E)
8	(9)	$P \vee R$	8,$\vee$I
8	(10)	$Q \vee (P \vee R)$	9,$\vee$I (ends 2nd inner proof)
5	(11)	$Q \vee (P \vee R)$	5,6,7,8,10,$\vee$E (ends 2nd outer proof)
1	(12)	$Q \vee (P \vee R)$	1,2,4,5,11,$\vee$E.

This proof suggests a procedural remark that parallels one made about CP. The remark was this: if a conditional is to be proved, consider using CP. The corresponding remark is: *if a sentence is to be proved from a disjunction, consider using $\vee$E.* The remark applies, even in the middle of a $\vee$E proof. At line 5 we assumed the second disjunct of the disjunction of line 1. Since at this stage we are in the middle of a $\vee$-elimination proof, the point of the assumption was to to have a second proof of the desired sentence from this, the second disjunct of line 1. But since this disjunct is itself a disjunction, it is natural to use a $\vee$E proof at this stage.

All of the previous three sequents used to illustrate $\vee$-elimination are natural deductive counterparts of axioms of *Principia Mathematica*. As a final example, we consider a fourth such case:

20 $P \vee P \vdash P$

1	(1)	$P \vee P$	A
2	(2)	P	A ($\vee$E)
1	(3)	P	1,2,2,2,2,$\vee$E.

This proof may be rather startling. But recall that an assumption in a proof is a proof of a sentence from itself; accordingly, the assumption of P at line 2 is a proof of P from P. In fact both of the proofs required for the $\vee$E application begin and end at line 2.

[handwritten: A Line justified by RAA must be a negation.]

2.2.12 Reductio ad Absurdum (RAA)

The final rule of the system corresponds to an ancient disputational device: that of defeating an opponent's position by demonstrating that it leads to an impossible or absurd or untenable conclusion. Its classical logical counterpart depends upon a proof of a *contradiction*, which for this system is defined as *a conjunction of a sentence with its negation in that order*. Thus P ∧ ¬P is a contradiction by this definition, but ¬P ∧ P is not, though evidently a contradiction can be proved from it. Specifically, the rule is:

A proof of a contradiction B ∧ ¬B from an ensemble Γ,A can be extended to a proof of ¬A from Γ.

To illustrate, consider

21 P → Q, P → ¬Q ⊢ ¬P

1	(1)	P → Q	A
2	(2)	P → ¬Q	A
3	(3)	P	A (RAA)
1,3	(4)	Q	1,3,MPP
2,3	(5)	¬Q	2,3,MPP
1,2,3	(6)	Q ∧ ¬Q	4,5,∧I
1,2	(7)	¬P	3,6,RAA.

[handwritten: Assume the opposite of the conclusion. and see if you produce this]

We have proved a contradiction (Q ∧ ¬Q) at line 6 from the ensemble of the sentences at lines 1,2, and 3. The rule permits us to write the negation of one assumption of the ensemble as proved from the remainder. Since we had set out to prove ¬P, P was the assumption we chose to negate.

Again, consider

22 P → ¬P ⊢ ¬P

1	(1)	P → ¬P	A
2	(2)	P	A (RAA)
1,2	(3)	¬P	1,2,MPP

$$1,2 \qquad (4) \quad P \wedge \neg P \qquad 2,3,\wedge I$$
$$1 \qquad (5) \quad \neg P \qquad 2,4,RAA$$

RAA takes a proof as its input; accordingly the justification requires that we cite the lines where the proof began, and where it ended, always a contradiction. On the extreme left of the RAA line we cite by line-number all of the assumptions from which the contradiction was proved *except the assumption negated at this line.*

RAA, as stated, is always used to prove a negation. However it can also play a role in proving an unnegated sentence A. In this case, assume ¬A; prove a contradiction; add the negation of the assumption (¬¬A) with RAA as the justification; then add A with DN as the justification.

Finally, a rule of thumb. When, in the course of constructing a proof, no other way of proceeding seems to present itself, try RAA.

2.2.13 Proofs and the nature of the rules

A logic student once came to me in some distress. She explained that she was sure that she could produce the required proofs if only she could figure out how to follow the rules. But, she went on, she couldn't figure out what the rules were 'trying to get her to do'. 'Anne,' I said, 'You should have been a nun.' 'I *was* a nun,' she replied. The point of the story is that the rules are to be understood as rules that *let* us do things, rather than rules that *require* us to do things. The constraint on a proof is that every line must be permitted by *some rule or other*. Only the set of all the rules constitutes a sort of check, since we may not write down a line that is not permitted by at least one of them. Each rule says 'If you already have a sequent with such and so features, then you *may* extend it in such and such a way.' Liberated by this disclosure, Anne went on to become a crackerjill sequent-prover.

Now having said all of that about the permissive character of the rules, one must add that it is possible to misapply them, for there are only precisely specified kinds of extensions that a given rule permits. For example, neither MPP nor any other rule of L will permit us to write down P citing only a previous line where we have written P → Q and a previous line where we have written Q. The system L lacks the rule (sometimes called 'Modus Ponendo Moron') that would permit that move. And again, no rule of L will permit us to write ¬Q, citing only a line where P → Q is written and a line where ¬P is written. Modus Tollendo Moron is not a rule of the system either.

2.2.14 The compactness and transitivity of ⊢

Let us remind ourselves what our understanding of an L-proof is:

> An L-proof of a sentence A from an ensemble Γ of sentences is a finite sequence of sentences, each of which is justified by a rule of L, and the last of which is A, having only sentences of Γ cited as assumptions.

What of the case in which Γ is infinite? In this case, only finitely many sentences from Γ can occur in the proof as assumptions, since the proof is a *finite* sequence. Let Γ' be the set of elements of Γ occurring in the proof. Evidently, there is a proof of A from that finite set. This property of provability is called *compactness*:

> If Γ ⊢ A, then for some finite subset Γ' of Γ, Γ' ⊢ A.

Finally, we can add to the two structural properties of classical provability (reflexivity and monotonicity) already noted, the third property of *transitivity*. To say that ⊢ is *transitive* is to say that

> if Γ,A ⊢ B and Γ ⊢ A, then Γ ⊢ B.

To see that this is a property of ⊢, assume that there is a proof of B from the ensemble Γ taken together with A. Such a proof is a finite sequence of sentences, which we can suppose has sentences of Γ as its first n entries all justified by the rule of assumption, then A as its next (n+1th) entry also justified by the rule of assumptions. The sequence continues until the entry B, at a line having only assumptions from Γ,A off to the left. But now suppose that there is a proof of A from Γ. We then replace line n+1 with the (renumbered) proof of A from Γ. Now the result of this replacement is that the line where A occurs will now cite at its extreme left a list of assumptions from Γ. For all of the later lines of the modified sequence of sentences, we renumber, and replace all citations of A on the left with citations of the assumptions from Γ from which A was proved. The resulting sequence of sentences will be a proof of B from Γ.

Exercise 2.3

Use **Simon**'s proof editor to construct a proof for each of the following sequents:

(a) $P \vdash Q \rightarrow (P \wedge Q)$

(b) $P \wedge (Q \wedge R) \vdash Q \wedge (P \wedge R)$

(c) $(P \rightarrow Q) \wedge (P \rightarrow R) \vdash P \rightarrow (Q \wedge R)$

(d) $Q \vdash P \vee Q$

(e) $P \wedge Q \vdash P \vee Q$

(f) $(P \rightarrow R) \wedge (Q \rightarrow R) \vdash (P \vee Q) \rightarrow R$

(g) $P \rightarrow Q, R \rightarrow S \vdash (P \wedge R) \rightarrow (Q \wedge S)$

(h) $P \rightarrow Q, R \rightarrow S \vdash (P \vee R) \rightarrow (Q \vee S)$

(i) $P \rightarrow (Q \wedge R) \vdash (P \rightarrow Q) \wedge (P \rightarrow R)$

(j) $\neg P \rightarrow P \vdash P$

2.2.15 The biconditional

The expression

$$A \leftrightarrow B$$

abbreviates the conjunction

$$(A \rightarrow B) \wedge (B \rightarrow A).$$

We read 'A $\leftrightarrow$ B', 'A if and only if B' or sometimes 'A ifif B'. The justification for the former of these readings lies in our understanding of the construction '. . . if and only if – – –' of natural language. A word or two on that subject is in order.

Circumstances may be such that you will arrive at a date on time *only if* you hurry. But that being so, it does not follow that *if* you hurry you will be on time. For all your hurrying, you may be struck down by the very bus you would otherwise have caught. A club may post a notice to the effect that one can join only if one is a woman; but it does not thereby commit itself to admitting every woman who applies. One might be permitted to join a club for retired professional women only if one is a woman, but one might be a woman (perhaps a vigorous, thrusting young business woman) and not be welcome as a member. The difference is that the sentence *You may join only if you are a woman* sets out a *necessary condition* for eligibility, whereas the sentence *You may join if you are a woman* sets out a *sufficient condition* for eligibility. Accordingly, to say *You may join if and only if you are a woman* is to give being a woman as both a necessary and a sufficient condition for eligibility. Both are expressible without recourse to the *only if* construction. The sufficiency of being a woman is expressed by *If you are a woman, then you are eligible*. The necessity of being a woman is expressed by *If you are eligible, then you are a woman*. Since the *only if* direction is represented by an arrow pointing from left to right, it is natural to represent the *if* direction by an arrow from right to left. The notation given combines the two. In technical writing, *if and only if* is sometimes written *iff*, which, like '$\leftrightarrow$', can be read 'ifif'.

The $\leftrightarrow$ is not primitive in our system; accordingly we provide no introduction or elimination rules for it. However, since it is used in abbreviations of sentences expressible in our primitive vocabulary, it can occur in sequents and proofs. Here sentences or parts of sentences can be replaced by equivalent primitive notation. Strictly speaking, such a replacement is not a (non-null) *extension* of a proof, since it results in the same entry in different notation, but it is a convenient convention to number the resulting sentence as a new line of the proof, with the output sentence justified by mention of the input sentence followed by 'df.$\leftrightarrow$' ('df.' for 'definition'). The practice is illustrated in the proofs of sequents expressed using $\leftrightarrow$.

23 $P \leftrightarrow Q \vdash Q \leftrightarrow P$ ($\leftrightarrow$-Commutation)

1	(1)	$P \leftrightarrow Q$	A
1	(2)	$(P \rightarrow Q) \wedge (Q \rightarrow P)$	1,df.$\leftrightarrow$
1	(3)	$P \rightarrow Q$	2,$\wedge$E
1	(4)	$Q \rightarrow P$	2,$\wedge$E
1	(5)	$(Q \rightarrow P) \wedge (P \rightarrow Q)$	3,4,$\wedge$I
1	(6)	$Q \leftrightarrow P$	5,df.$\leftrightarrow$

24 $P, P \leftrightarrow Q \vdash Q$

1	(1)	P	A
2	(2)	$P \leftrightarrow Q$	A
2	(3)	$(P \rightarrow Q) \wedge (Q \rightarrow P)$	2,df.$\leftrightarrow$
2	(4)	$P \rightarrow Q$	3,$\wedge$E
1,2	(5)	Q	1,4,MPP

25 $P \leftrightarrow Q, Q \leftrightarrow R \vdash P \leftrightarrow R$

1	(1)	$P \leftrightarrow Q$	A
2	(2)	$Q \leftrightarrow R$	A
1	(3)	$(P \rightarrow Q) \wedge (Q \rightarrow P)$	1,df.$\leftrightarrow$
2	(4)	$(Q \rightarrow R) \wedge (R \rightarrow Q)$	2,df.$\leftrightarrow$
1	(5)	$P \rightarrow Q$	3,$\wedge$E
2	(6)	$Q \rightarrow R$	4,$\wedge$E
7	(7)	P	A (CP)
1,7	(8)	Q	5,7,MPP
1,2,7	(9)	R	6,8,MPP
1,2	(10)	$P \rightarrow R$	7,9,CP
1	(11)	$Q \rightarrow P$	3,$\wedge$E
2	(12)	$R \rightarrow Q$	4,$\wedge$E
13	(13)	R	A (CP)
2,13	(14)	Q	12,13,MPP
1,2,13	(15)	P	11,14,MPP
1,2	(16)	$R \rightarrow P$	13,15,CP
1,2	(17)	$(P \rightarrow R) \wedge (R \rightarrow P)$	10,16,$\wedge$I
1,2	(18)	$P \leftrightarrow R$	17,df.$\leftrightarrow$

$(P \supset R) \wedge (R \supset P)$

26    $(P \wedge Q) \leftrightarrow P \vdash P \to Q$

1	(1)	$(P \wedge Q) \leftrightarrow P$	A
1	(2)	$((P \wedge Q) \to P) \wedge (P \to (P \wedge Q))$	1,Df.$\leftrightarrow$
1	(3)	$(P \to (P \wedge Q))$	2,$\wedge$E
4	(4)	P	A (CP)
1,4	(5)	$P \wedge Q$	3,4,MPP
1,4	(6)	Q	5,$\wedge$E
1	(7)	$P \to Q$	4,6,CP

27    $P \wedge (P \leftrightarrow Q) \vdash P \wedge Q$

1	(1)	$P \wedge (P \leftrightarrow Q)$	A
1	(2)	$P \wedge ((P \to Q) \wedge (Q \to P))$	1,df.$\leftrightarrow$
1	(3)	P	2,$\wedge$E
1	(4)	$(P \to Q) \wedge (Q \to P)$	2,$\wedge$E
1	(5)	$P \to Q$	4,$\wedge$E
1	(6)	Q	3,5,MPP
1	(7)	$P \wedge Q$	3,6,$\wedge$I

Exercise 2.4

Use **Simon**'s proof editor to construct a proof for each of the following sequents

(a) $Q, P \leftrightarrow Q \vdash P$

(b) $\neg Q, P \leftrightarrow Q \vdash \neg P$

(c) $P \to Q, Q \to P \vdash P \leftrightarrow Q$

(d) $P \leftrightarrow Q \vdash \neg P \leftrightarrow \neg Q$

(e) $\neg P \leftrightarrow \neg Q \vdash P \leftrightarrow Q$

(f) $(P \vee Q) \leftrightarrow P \vdash Q \to P$

(g) $P \leftrightarrow \neg Q, Q \leftrightarrow \neg R \vdash P \leftrightarrow R$

2.2.16 Some useful sequents and their proofs

There is no better way to acquire facility in constructing proofs than to combine a careful study of proofs already worked out with plenty of practice in constructing one's own. That in itself would justify setting out plenty of sample proofs. For the student advancing to more logical studies, however, there is a great advantage in having a broad and easy familiarity with the connectives of classical propositional logic, and particularly with their formal inferential properties. For it is often useful to be able to recast sentences into equivalent forms, and therefore to know without mental effort what can be proved from what. There are, of course, unsolved proof-theoretic problems, even within classical propositional logic, and new proof-theoretic techniques as yet undiscovered. No one knows all there is to be known about what can be proved from what.

Study each of the sequents and proofs that follow, both to observe what can be proved from what, and to understand the technique by which it is proved. Even the very simple is worthy of scrutiny. Consider the earlier claim that the rule assumption permits the proof of a sentence from itself. Here is that proof:

28 $P \vdash P$

1   (1)   P   A.

Notice that this satisfies the conditions for being an L-proof. It is a finite sequence of sentences. Every sentence in the sequence is justified by a rule of L. The last sentence is the sentence we were to prove, and the only assumption listed to the left of it is the premiss from which it was to be proved.

$A \dashv\vdash B$ abbreviates the claim $A \vdash B$ and $B \vdash A$. Such a pair of sentences are said to be *interderivable* and *deductively equivalent*. The former because each can be proved from the other, and the latter because it follows that whatever can be proved from the one can be proved from the other. Evidently sequents 8 and 11 taken together prove the deductive equivalence of $(P \wedge Q) \rightarrow R$ and $P \rightarrow (Q \rightarrow R)$. Thus:

29 $(P \wedge Q) \rightarrow R \dashv\vdash P \rightarrow (Q \rightarrow R)$

Proofs of deductive equivalence require both a left-to-right and a right-to-left proof labelled as in the following examples.

30 $P \wedge (P \vee Q) \dashv\vdash P$

| ($\vdash$) | 1 | (1) | $P \wedge (P \vee Q)$ | A |
| | 1 | (2) | P | 1,$\wedge$E |

($\dashv$)	1	(1)	P	A
	1	(2)	$P \vee Q$	1,$\vee$I
	1	(3)	$P \wedge (P \vee Q)$	1,2,$\wedge$I

31 $P \vee (P \wedge Q) \dashv\vdash P$

($\vdash$)	1	(1)	$P \vee (P \wedge Q)$	A
	2	(2)	P	A ($\vee$E)
	3	(3)	$P \wedge Q$	A ($\vee$E)
	3	(4)	P	3,$\wedge$E
	1	(5)	P	1,2,2,3,4,$\vee$E

| ($\dashv$) | 1 | (1) | P | A |
| | 1 | (2) | $P \vee (P \wedge Q)$ | 1,$\vee$I |

32 $P \to Q, \neg Q \vdash \neg P$ (Modus Tollendo Tollens (MTT))

	1	(1)	$P \to Q$	A
	2	(2)	$\neg Q$	A
	3	(3)	P	A (RAA)
	1,3	(4)	Q	1,3,MPP
	1,2,3	(5)	$Q \wedge \neg Q$	2,4,$\wedge$I
	1,2	(6)	$\neg P$	3,5,RAA

33 $P \to \neg Q, Q \vdash \neg P$

	1	(1)	$P \to \neg Q$	A
	2	(2)	Q	A
	3	(3)	P	A (RAA)
	1,3	(4)	$\neg Q$	1,3,MPP
	1,2,3	(5)	$Q \wedge \neg Q$	2,4,$\wedge$I
	1,2	(6)	$\neg P$	3,5,RAA

34 $\neg P \to Q, \neg Q \vdash P$

	1	(1)	$\neg P \to Q$	A
	2	(2)	$\neg Q$	A
	3	(3)	$\neg P$	A (RAA)
	1,3	(4)	Q	1,3,MPP

1,2,3	(5)	Q ∧ ¬Q	2,4,∧I
1,2	(6)	¬¬P	3,5,RAA
1,2	(7)	P	6,DN

35 P → Q, Q → R, ¬R ⊢ ¬P

1	(1)	P → Q	A
2	(2)	Q → R	A
3	(3)	¬R	A
4	(4)	P	A (RAA)
1,4	(5)	Q	1,4,MPP
1,2,4	(6)	R	2,5,MPP
1,2,3,4	(7)	R ∧ ¬R	3,6,∧I
1,2,3	(8)	¬P	4,7,RAA

36 P → (Q → R), P, ¬R ⊢ ¬Q

1	(1)	P → (Q → R)	A
2	(2)	P	A
3	(3)	¬R	A
4	(4)	Q	A (RAA)
1,2	(5)	Q → R	1,2,MPP
1,2,4	(6)	R	4,5,MPP
1,2,3,4	(7)	R ∧ ¬R	3,6,∧I
1,2,3	(8)	¬Q	4,7,RAA

37 P → Q ⊢ ¬Q → ¬P

1	(1)	P → Q	A
2	(2)	¬Q	A (CP)
3	(3)	P	A (RAA)
1,3	(4)	Q	1,3,MPP
1,2,3	(5)	Q ∧ ¬Q	2,4,∧I
1,2	(6)	¬P	3,5,RAA
1	(7)	¬Q → ¬P	2,6,CP

38 Q → R ⊢ (¬Q → ¬P) → (P → R)

1	(1)	Q → R	A
2	(2)	¬Q → ¬P	A (CP)
3	(3)	P	A (CP)
4	(4)	¬Q	A (RAA)
2,4	(5)	¬P	2,4,MPP
2,3,4	(6)	P ∧ ¬P	3,5,∧I

2,3	(7)	¬¬Q	4,6,RAA
2,3	(8)	Q	7,DN
1,2,3	(9)	R	1,8,MPP
1,2	(10)	P → R	3,9,CP
1	(11)	(¬Q → ¬P) → (P → R)	2,10,CP

39   P, ¬(P ∧ Q) ⊢ ¬Q (Modus Ponendo Tollens (MPT))

1	(1)	P	A
2	(2)	¬(P ∧ Q)	A
3	(3)	Q	A (RAA)
1,3	(4)	P ∧ Q	1,3,∧I
1,2,3	(5)	(P ∧ Q) ∧ ¬(P ∧ Q)	2,4,∧I
1,2	(6)	¬Q	3,5,RAA

40   P → Q ⊣⊢ ¬(P ∧ ¬Q)

(⊢)	1	(1)	P → Q	A
	2	(2)	P ∧ ¬Q	A (RAA)
	2	(3)	P	2,∧E
	2	(4)	¬Q	2,∧E
	1,2	(5)	Q	1,3,MPP
	1,2	(6)	Q ∧ ¬Q	4,5,∧I
	1	(7)	¬(P ∧ ¬Q)	2,6,RAA

(⊣)	1	(1)	¬(P ∧ ¬Q)	A
	2	(2)	P	A (CP)
	3	(3)	¬Q	A (RAA)
	2,3	(4)	P ∧ ¬Q	2,3,∧I
	1,,2,3	(5)	(P ∧ ¬Q) ∧ ¬(P ∧ ¬Q)	1,4,∧I
	1,2	(6)	¬¬Q	3,5,RAA
	1,2	(7)	Q	6,DN
	1	(8)	P → Q	2,7,CP

41   P ∨ Q ⊣⊢ ¬(¬P ∧ ¬Q)

(⊢)	1	(1)	P ∨ Q	A
	2	(2)	¬P ∧ ¬Q	A (RAA)
	3	(3)	P	A (∨E)
	2	(4)	¬P	2,∧E
	2,3	(5)	P ∧ ¬P	3,4,∧I

3	(6)	$\neg(\neg P \wedge \neg Q)$	2,5,RAA
7	(7)	Q	A ($\vee$E)
2	(8)	$\neg Q$	2,$\wedge$E
2,7	(9)	$Q \wedge \neg Q$	7,8,$\wedge$I
7	(10)	$\neg(\neg P \wedge \neg Q)$	2,9,RAA
1	(11)	$\neg(\neg P \wedge \neg Q)$	1,3,6,7,10,$\vee$E

$\neg(\neg P \wedge \neg Q) \vdash P \vee Q$ $\vdash P \vee Q$

(⊣) 1	(1)	$\neg(\neg P \wedge \neg Q)$	A
2	(2)	$\neg(P \vee Q)$	A (RAA)
3	(3)	P	A (RAA)
3	(4)	$P \vee Q$	3,$\vee$I
2,3	(5)	$(P \vee Q) \wedge \neg(P \vee Q)$	2,4,$\wedge$I
2	(6)	$\neg P$	3,5,RAA
7	(7)	Q	A (RAA)
7	(8)	$P \vee Q$	7,$\vee$I
2,7	(9)	$(P \vee Q) \wedge \neg(P \vee Q)$	2,8,$\wedge$I
2	(10)	$\neg Q$	7,9,RAA
2	(11)	$\neg P \wedge \neg Q$	6,10,$\wedge$I
1,2	(12)	$(\neg P \wedge \neg Q) \wedge \neg(\neg P \wedge \neg Q)$	1,11,$\wedge$I
1	(13)	$\neg\neg(P \vee Q)$	2,12,RAA
1	(14)	$P \vee Q$	13,DN

The following pair of deductive equivalents are usually referred to erroneously as DeMorgan's theorems, after the mathematician, Augustus DeMorgan (1806-1871). In essentially the form in which they are presented here, they were known to Petrus Hispanus (Peter of Spain, later Pope John XXI) (d.1277), and certainly to earlier logicians.

42 $\neg(P \wedge Q) \dashv\vdash \neg P \vee \neg Q$ (DeM)

($\vdash$) 1	(1)	$\neg(P \wedge Q)$	A
2	(2)	$\neg(\neg P \vee \neg Q)$	A (RAA)
3	(3)	$\neg P$	A (RAA)
3	(4)	$\neg P \vee \neg Q$	3,$\vee$I
2,3	(5)	$(\neg P \vee \neg Q) \wedge \neg(\neg P \vee \neg Q)$	2,4,$\wedge$I
2	(6)	$\neg\neg P$	3,5,RAA
2	(7)	P	6,DN
8	(8)	$\neg Q$	A (RAA)
8	(9)	$\neg P \vee \neg Q$	8,$\vee$I
2,8	(10)	$(\neg P \vee \neg Q) \wedge \neg(\neg P \vee \neg Q)$	2,9,$\wedge$I

2	(11)	¬¬Q	8,10,RAA
2	(12)	Q	11,DN
2	(13)	P ∧ Q	7,12,∧I
1,2	(14)	(P ∧ Q) ∧ ¬(P ∧ Q)	2,13,∧I
1	(15)	¬¬(¬P ∨ ¬Q)	2,14,RAA
1	(16)	¬P ∨ ¬Q	15,DN

¬P ∨ ¬Q ⊢ ¬(P ∧ Q)

(⊣) 1	(1)	¬P ∨ ¬Q	A
2	(2)	P ∧ Q	A (RAA)
3	(3)	¬P	A (∨E)
2	(4)	P	2,∧E
2,3	(5)	P ∧ ¬P	3,4,∧I
3	(6)	¬(P ∧ Q)	2,5,RAA
7	(7)	¬Q	A (∨E)
2	(8)	Q	2,∧E
2,7	(9)	Q ∧ ¬Q	7,8,∧I
7	(10)	¬(P ∧ Q)	2,9,RAA
1	(11)	¬(P ∧ Q)	1,3,6,7,10,∨E

43   ¬(P ∨ Q) ⊣⊢ ¬P ∧ ¬Q (DeM)

(⊢) 1	(1)	¬(P ∨ Q)	A
2	(2)	P	A (RAA)
2	(3)	P ∨ Q	2,∨I
1,2	(4)	(P ∨ Q) ∧ ¬(P ∨ Q)	1,3,∧I
1	(5)	¬P	2,4,RAA
6	(6)	Q	A (RAA)
6	(7)	P ∨ Q	6,∨I
1,6	(8)	(P ∨ Q) ∧ ¬(P ∨ Q)	1,7,∧I
1	(9)	¬Q	6,8,RAA
1	(10)	¬P ∧ ¬Q	5,9,∧I

¬P ∧ ¬Q ⊢ ¬(P ∨ Q)

The goal is to create the contradiction of the premiss

(⊢) 1	(1)	¬P ∧ ¬Q	A
2	(2)	P ∨ Q	A (RAA)
3	(3)	P	A (∨E)
1	(4)	¬P	1,∧E
1,3	(5)	P ∧ ¬P	3,4,∧I
3	(6)	¬(¬P ∧ ¬Q)	1,5,RAA
7	(7)	Q	A (∨E)
1	(8)	¬Q	1,∧E

New assumption that wasn't put in as A CRAA of an assumption

1,7	(9)	$Q \wedge \neg Q$	7,8,$\wedge$I
7	(10)	$\neg(\neg P \wedge \neg Q)$	1,9,RAA
2	(11)	$\neg(\neg P \wedge \neg Q)$	2,3,6,7,10,$\vee$E
1,2	(12)	$(\neg P \wedge \neg Q) \wedge \neg(\neg P \wedge \neg Q)$	1,11,$\wedge$I
1	(13)	$\neg(P \vee Q)$	2,12,RAA

Exercise 2.5

Use **Simon**'s proof editor to construct a proof for each of the following sequents.

(a) $\neg\neg Q \rightarrow P, \neg P \vdash \neg Q$
(b) $\neg P \rightarrow \neg Q, Q \vdash P$
(c) $P \rightarrow \neg Q \vdash Q \rightarrow \neg P$
(d) $\neg P \rightarrow Q \vdash \neg Q \rightarrow P$
(e) $\neg P \rightarrow \neg Q \vdash Q \rightarrow P$
(f) $P \vdash (\neg(Q \rightarrow R) \rightarrow \neg P) \rightarrow (\neg R \rightarrow \neg Q)$

2.2.17 Summary of the Rules of L

Here, for easy reference, the nine primitive rules of L are summarized along with relevant practical information related to their use and documentation. The assumptions are always specified by line number

The Rule of Assumption (A)

Rule: Any sentence can be written as a line of a proof.

Entry: any sentence
Justification:A
Required assumptions: the sentence at this line.

Modus Ponendo Ponens (MPP)

Rule: A proof that includes a proof of A $\rightarrow$ B from the ensemble Σ, and a proof of A from the ensemble Γ, can be extended to a proof of B from the ensemble Σ,Γ.

Entry: B

Justification: <line no. of A → B>,<line no. of A>,MPP
Required assumptions: those of A → B and A

The Rule of Double Negation (DN)

Rule: A proof of a sentence ¬¬A from an ensemble Σ can be extended to
a proof of A from the ensemble Σ.

Entry: A
Justification: <line no. of ¬¬A>,DN
Required assumptions: those of ¬¬A

The Rule of Conditional Proof (CP)

Rule: A proof of B from an ensemble Σ,A can be extended to a proof of
A → B from Σ.

Entry: A → B
Justification: <line no. of A>,<line no. of B>,CP
Required assumptions: those of B, except A

The Rule of ∧-introduction (∧I)

Rule: A proof that includes both a proof of A from ensemble Γ, and a
proof of B from ensemble Σ can be extended to a proof of A ∧ B
from the ensemble Γ,Σ.

Entry: A ∧ B
Justification: <line no. of A>,<line no. of B>,∧I
Required assumptions: those of A and B

The Rule of ∧-elimination (∧E)

Rule: A proof of A ∧ B from ensemble Γ can be extended to a proof of A
(B) from Γ.

Entry: A (B)

Justification: <line no. of A ∧ B>,∧E
Required assumptions: those of A ∧ B

The Rule of ∨-Introduction (∨I)

Rule: A proof of A from ensemble Γ can be extended to a proof of A ∨ B
(B ∨ A) from the ensemble Γ.

Entry: A ∨ B (B ∨ A)
Justification: <line no. of A>,∨I
Required assumptions: those of A

The Rule of ∨-Elimination (∨E)

Rule: A proof that includes (a) a proof of a disjunction A ∨ B from
ensemble Γ (b) a proof of C from the ensemble Δ,A and (c) a proof
of C from ensemble Σ,B can be extended to a proof of C from
Γ,Δ,Σ.

Entry: C
Justification: <line no. of A ∨ B>,<line no. of A>,<line no. of C (proved
from A)>,<line no. of B>,<line no. of C (proved from B)>,∨E
Required assumptions: Those of A ∨ B, those of C (proved from A)
except A, those of C (proved from B) except B.

Reductio Ad Absurdum (RAA)

Rule: A proof of a contradiction B ∧ ¬B from an ensemble Γ,A can be
extended to a proof of ¬A from Γ.

Entry: ¬A
Justification: <line no. of A>, <line no. of B ∧ ¬B>,RAA

Exercise 2.6

Use **Simon**'s proof editor to construct a proof for each of the following sequents.

(a) $P \vee Q \vdash P \vee Q$

(b) $P \wedge P \dashv\vdash P$

(c) $P \wedge (Q \vee R) \dashv\vdash (P \wedge Q) \vee (P \wedge R)$

(d) $P \vee (Q \wedge R) \dashv\vdash (P \vee Q) \wedge (P \vee R)$

(e) $P \wedge Q \dashv\vdash \neg(P \rightarrow \neg Q)$

(f) $P \wedge Q \dashv\vdash \neg(\neg P \vee \neg Q)$

(g) $P \rightarrow Q \vdash \neg P \vee Q$

(h) $\neg P \rightarrow Q \vdash P \vee Q$

3

Propositional Logic 2

Introduction

Our approach so far has been an informal one. We have used the terms *language, formal system,* and *proof,* but we have not disclosed officially what these objects are. Soon, however, we will want to demonstrate some theorems (called *metatheorems*) about our formal system, and in doing so we will be helped by precise definitions of the objects of study.

3.1 Formal systems

A *formal system $\underline{S}$ of natural deduction* is an ordered pair[1]

$$<L, R>$$

where L is a language and R is a set of rules. R, with one minor change of wording, is just the set of rules with which we are now familiar. The language (said to be the underlying language of the system S) is just the language in which we have been constructing proofs, but here we must give a more precise account of it.

---

1. For present purposes, an ordered pair (triple, quadruple,. . . n-tuple) is a set of 2 (3, 4, . . .n) elements of which the order is an essential feature. Typically in logical settings, the items specified earlier provide the means for the specifications of the later items. In general, an ordered set for which identity is not preserved by permutations of its elements. So, for example, the ordered set <a, b> is not identical to the ordered set <b, a>, and if an ordered set <a, b> is identical to an ordered set <c, d>, then a is identical to c, and b is identical to d.

3.1.1 The language

In any formal system S, the underlying language, L of S is itself an ordered triple:

$$<At, k, \Phi>$$

where:

At is a set of *atoms*
k is a set of constants
Φ is a set of well-formed formulae (wffs), those formulae that are correctly constructed in accordance with the grammar of the language.

What follows is an account of the language underlying the particular propositional system L that we have been studying.

At is a denumerable[2] set of symbols called *propositional variables*:

$$P_0, P_1, \ldots, P_i, \ldots$$

though for convenience we will continue to use

$$P, Q, R, \ldots$$

k is the set of symbols:

'(', ')' (left and right hand *brackets*)

'$\rightarrow$', '$\neg$', '$\wedge$', '$\vee$' (the *primitive connectives*).

To define Φ, the set of well-formed formulae, we will once again have recourse to a set of variables, (A, B, C, A_1, A_2,... and so on, called *metalogical variables*, ranging over *formulae*. The letter 'A' names an arbitrary wff. First recall that by *symbol* is meant a bracket, a

---

2. A denumerable set is a set the elements of which can be put in a one-to-one correspondence with the set of counting numbers (0, 1, 2, ...)

propositional variable or a primitive connective. Any finite string of
symbols is a *formula*. So, for example,

$$\rightarrow P\neg\wedge QRS\vee$$

is a formula. What we must now do is say which formulae are well-
formed. We do this by an *inductive definition*. The *basis* of the definition
is the clause defining the shortest wffs (shortest in number of occurrences
of symbols). Additional clauses then tell us for wffs of arbitrary finite
length, how they can be combined to form new wffs. Finally an *extremal
clause* tells us that nothing else is a wff.

1. Any atomic sentence (propositional variable) is a wff.
2. If A is a wff, then $\neg$A is a wff.
3. If A is a wff and B is a wff, then (A $\rightarrow$ B) is a wff.
4. If A is a wff and B is a wff, then (A $\vee$ B) is a wff.
5. If A is a wff and B is a wff, then (A $\wedge$ B) is a wff.
6. If A is not a wff in virtue of 1 – 5, then A is not a wff.

An equivalent alternative account of Φ is given in terms of the notion of
closure. To say that a set A is *closed* under an n-ary operation * is to say
that if $x_1, x_2, \ldots x_n$ are all members of A, then *$(x_1, x_2, \ldots x_n)$ is a
member of the set. As an example, the set of natural numbers is closed
under multiplication and addition, but not under division.

Φ can be defined as the smallest set of formulae which (a) includes *At*
and (b) is *closed* under negation, conditionalization, conjunction and
disjunction.

That is a sufficient official characterization of the underlying language of
L. It is the definition to which we must defer when we are proving
metatheorems about all of the sentences of Φ. However, for day-to-day
work in the system, we can simplify things conventionally by the
introduction of some shortcuts. First, we introduce the connective $\leftrightarrow$ as
an abbreviation by the definition

$$A \leftrightarrow B =_{df} (A \rightarrow B) \wedge (B \rightarrow A).$$

Second, we simplify the way in which we write wffs by the convention
that *all outermost brackets may be omitted*. That is, we will write

$$P \lor (Q \land R)$$

rather than

$$(P \lor (Q \land R)).$$

And again, to reduce the overburden of bracketting, the connectives of our language will be understood to have differing *binding powers*. These are given in decreasing order of strength by the following conventions:

1. '$\neg$' $\neg P \lor Q$ is understood as $(\neg P) \lor Q$.
2. '$\lor$', '$\land$' $P \land Q \to R$ is understood as $(P \land Q) \to R$.
3. '$\to$' $P \to Q \leftrightarrow R$ is understood as $(P \to Q) \leftrightarrow R$.
4. '$\leftrightarrow$'

Note that $P \land Q \lor R$ is not well-formed; were it permitted, it would be ambiguous as between:

$$(P \land Q) \lor R \text{ and } P \land (Q \lor R)$$

and those two wffs must therefore be distinguished by bracketting.

The binding power of a connective is inversely proportional to its *scope*. The *scope* of an occurrence of a connective is the smallest wff of which that occurrence is a part. Thus, in $P \leftrightarrow Q \to R$, the scope of the occurrence of $\leftrightarrow$ is the whole wff; the scope of the occurrence of $\to$ is the wff $Q \to R$. So where brackets do not make the matter explicit, $\leftrightarrow$ has the longest scope, $\to$ the next longest, $\land$ and $\lor$ the next, and $\neg$ the shortest scope.

'$\vdash$' (the assertion sign, or 'turnstile') and '$\dashv\vdash$' (the double turnstile) are further *metalogical symbols*.

a finite sequence of wffs followed by '$\vdash$' followed by a single wff

$$A_1, A_2, \ldots, A_n \vdash B$$

is a called a *sequent*. We have now officially given ourselves all of the notation that we used in the previous chapter. We can now adopt an official account of an L-proof.

An L-proof of a wff A from an ensemble Γ of wffs is a finite sequence of wffs, $B_1, \ldots, B_n$, where each B_i ($1 \le i \le n$) is justified by a rule of L, and where B_n is A and requires as assumptions only wffs in Γ.

3.2 Theorems

Consider the special case in which $\Gamma \vdash A$ and $\Gamma = \varnothing$. In this case, the proof of Γ will be a finite sequence of wffs, every one of which is justified by a rule of L, but the last of which is A requiring no assumptions at all. We abbreviate $\varnothing \vdash A$ ($\varnothing \vdash_L A$) by $\vdash A$ ($\vdash_L A$), which we read as 'A is a theorem of L' or 'A is an L-theorem.'

An L-theorem is a wff that is L-provable from the empty ensemble of assumptions.

The following are examples of theorems and proofs.

44 $\vdash \neg(P \wedge \neg P)$ (The Law of Non-contradiction)

1	(1)	$P \wedge \neg P$	A (RAA)
	(2)	$\neg(P \wedge \neg P)$	1,1,RAA

45 $\vdash P \rightarrow P$ (The Law of Identity)

1	(1)	P	A (CP)
	(2)	$P \rightarrow P$	1,1,CP

46 $\vdash P \rightarrow \neg\neg P$

1	(1)	P	A (CP)
2	(2)	$\neg P$	A (RAA)
1,2	(3)	$P \wedge \neg P$	1,2,$\wedge$I
1	(4)	$\neg\neg P$	2,3,RAA
	(5)	$P \rightarrow \neg\neg P$	1,4,CP

47 $\vdash \neg\neg P \rightarrow P$

48 $\vdash P \wedge Q \rightarrow P$

49 $\vdash (P \rightarrow Q) \rightarrow (\neg Q \rightarrow \neg P)$

1	(1)	$P \rightarrow Q$	A (CP)
2	(2)	$\neg Q$	A (CP)
3	(3)	P	A (RAA)
1,3	(4)	Q	1,3,MPP
1,2,3	(5)	$Q \wedge \neg Q$	2,4,∧I
1,2	(6)	$\neg P$	3,5,RAA
1	(7)	$\neg Q \rightarrow \neg P$	2,6,CP
	(8)	$(P \rightarrow Q) \rightarrow (\neg Q \rightarrow \neg P)$	1,7,CP

50 $\vdash (P \rightarrow (Q \rightarrow R)) \rightarrow ((P \rightarrow Q) \rightarrow (P \rightarrow R))$

1	(1)	$P \rightarrow (Q \rightarrow R)$	A (CP)
2	(2)	$P \rightarrow Q$	A (CP)
3	(3)	P	A (CP)
1,3	(4)	$Q \rightarrow R$	1,3,MPP
2,3	(5)	Q	2,3,MPP
1,2,3	(6)	R	4,5,MPP
1,2	(7)	$P \rightarrow R$	3,6,CP
1	(8)	$(P \rightarrow Q) \rightarrow (P \rightarrow R)$	2,7,CP
	(9)	$P \rightarrow (Q \rightarrow R) \rightarrow ((P \rightarrow Q) \rightarrow (P \rightarrow R))$	1,8,CP

51 $\vdash P \vee \neg P$ (The Law of Excluded Middle LEM (*Tertium Non Datur*))

1	(1)	$\neg(P \vee \neg P)$	A (RAA)
2	(2)	P	A (RAA)
2	(3)	$P \vee \neg P$	2,∨I
1,2	(4)	$(P \vee \neg P) \wedge \neg(P \vee \neg P)$	1,3,∧I
1	(5)	$\neg P$	2,4,RAA
1	(6)	$P \vee \neg P$	5,∨I
1	(7)	$(P \vee \neg P) \wedge \neg(P \vee \neg P)$	1,6,∧I
	(8)	$\neg\neg(P \vee \neg P)$	1,7,RAA
	(9)	$P \vee \neg P$	8,DN

Both of the tags attached to the last theorem are ancient. The first is intended to convey the notion that between a sentence and its negation, there is no distinct middle case. The intent of the Latin tag (A third is not given) is the same.

3.3 Uniform Substitution

Consider the last proof of the previous section. In it we prove that for the particular atom P, the wff that disjoins P with ¬P is a theorem. Evidently, we could have chosen the atom Q or any other atomic wff and constructed an otherwise identical proof. We could moreover have argued in the metalanguage in the following way:

Let A be any wff, then the following will be a proof:

1	(1)	¬(A ∨ ¬A)	A (RAA)
2	(2)	A	A (RAA)
2	(3)	A ∨ ¬A	2,∨I
1,2	(4)	(A ∨ ¬A) ∧ ¬(A ∨ ¬A)	1,3,∧I
1	(5)	¬A	2,4,RAA
1	(6)	A ∨ ¬A	5,∨I
1	(7)	(A ∨ ¬A) ∧ ¬(A ∨ ¬A)	1,6,∧I
	(8)	¬¬(A ∨ ¬A)	1,7,RAA
	(9)	A ∨ ¬A	8,DN.

We would thus have proved that for any wff A, the disjunction of A with its own negation is a theorem of L. Knowing that this is so, not just for the above theorem, but for any theorem whatsoever, we can avoid the necessity of constructing a new proof for each distinct wff by recognizing this universal feature of the system: that each L-theorem is also a representative of infinitely many others. To exploit this fact we introduce a notion of *uniform substitution.*

A wff A is said to have been obtained from a wff B by *uniform-substitution* when A is the result of substituting an occurrence of a wff C for *every* occurrence of some *atom* in B.

The substitution might result in all occurrences of one atom being replaced by occurrences of some other atom, as, say, all occurrences of P being replaced by occurrences of Q. We can also recognize as an instance of uniform substitution the case in which a propositional variable is replaced by itself (the vacuous substitution).

If the wff A is obtained from B as the result of a sequence of uniform substitutions, then A is said to be a *substitution instance* of B.

As an example, let B be the wff

$$(P \rightarrow Q) \rightarrow (\neg Q \rightarrow \neg P).$$

Let C be the wff (R → S). Then the wff

$$((R \rightarrow S) \rightarrow Q) \rightarrow (\neg Q \rightarrow \neg(R \rightarrow S))$$

is the result of uniformly substituting an occurrence of C for every occurrence of P in B. Alternatively, let C be the wff Q. Then the wff

$$(Q \rightarrow Q) \rightarrow (\neg Q \rightarrow \neg Q)$$

will result. Again, let C be the atom R. Then uniform substitution in B of C for Q yields

$$(P \rightarrow R) \rightarrow (\neg R \rightarrow \neg P).$$

Now substitute Q uniformly for P to obtain

$$(Q \rightarrow R) \rightarrow (\neg R \rightarrow \neg Q)$$

and finally substitute P uniformly for R:

$$(Q \rightarrow P) \rightarrow (\neg P \rightarrow \neg Q).$$

We now have a substitution instance of B in which the positions of P and Q are interchanged.

Remember that to count as a uniform substitution the substitution must satisfy two requirements. Any wff may be substituted, but the substitution must be made (a) *for an atom* and (b) *for every occurrence of it*. This last is the force of the word 'uniform'. The restrictions are important. All non-vacuous uniform substitutions for atoms within the scope of negation will result in some new negation, but we may not make arbitrary substitutions for negated atoms. From

$$(P \rightarrow Q) \rightarrow (\neg Q \rightarrow \neg P).$$

we do not obtain

$$(P \rightarrow Q) \rightarrow (Q \rightarrow P)$$

by uniform substitution. Now with that understanding of uniform substitution, we are in a position to state our first metatheorem about the system L.

Metatheorem 3.0

If a wff A is L-provable from $\varnothing$, then any wff B, obtainable by uniform substitution in A, is L-provable from $\varnothing$. Alternatively, the set of L-theorems is closed under uniform substitution.

The definition of uniform substitution can be extended to apply to all sequents. A sequent S' is said to have been obtained from a sequent expression S by uniform substitution iff S' is the result of substituting some wff A for every occurrence of some atom in S. As in the case of wffs, any sequent can be obtained from itself by uniform substitution.

Evidently, what we observed about theorems applies to arbitrary sequents. If the sequent $\Sigma' \vdash A'$ is the result of the uniform substitution of the wff B for P_i in the sequent $\Sigma \vdash A$, then if $\Sigma \vdash A$ has a proof, then a proof of $\Sigma' \vdash A'$ is obtained by uniformly substituting B for P_i throughout the proof of $\Sigma \vdash A$. Thus,

Metatheorem 3.1

The set of provable sequents is closed under uniform substitution.

3.4 Theorem and Sequent Introduction

3.4.1 The Rule of Theorem Introduction (TI)

Suppose that in the course of constructing a proof for a sequent, we realize that the remainder of the proof would be straightforward if only we had P ∨ ¬P as an entry. Clearly, we could, at that point in the proof, simply interpolate a proof of P ∨ ¬P without the addition of any new assumptions, and the same would be true for any wff obtainable from P ∨ ¬P by uniform substitution. But since we already have a proof of P ∨ ¬P, the additional lines of the proof need only reproduce the lines of a proof that we already have, or at any rate lines that could be obtained from the lines of that proof by uniform substitution. In the circumstances, there is hardly any point in requiring that the original proof or a simulacrum of it be reproduced. Accordingly we adopt as a *derived* (that is as a *non-primitive*) rule of L, a rule for the introduction of theorems:

The Rule of Theorem Introduction (TI)

Any substitution instance of any theorem already proved may be introduced as a line of a proof.

The entry is justified by 'TI' together with an indication of where its proof is to be found. If the theorem introduced is obtained by uniform substitution from the one being cited, then it is justified by 'TI(S)' together with an indication of where it is proved.

A note about Simon

Simon will admit the use of TI or TI(S) only for those theorems for which it finds a proof in your **Simon** document.

As an example of the use and the utility of TI, consider the sequent

$$P \rightarrow Q, \neg P \rightarrow R \vdash Q \vee R.$$

In the absence of TI, perhaps the most natural proof for it is the following:

1	(1)	$P \rightarrow Q$	A
2	(2)	$\neg P \rightarrow R$	A
3	(3)	$\neg(Q \vee R)$	A (RAA)
4	(4)	Q	A (RAA)
4	(5)	$Q \vee R$	4,$\vee$I
3,4	(6)	$(Q \vee R) \wedge \neg(Q \vee R)$	3,5,$\wedge$I
3	(7)	$\neg Q$	4,6,RAA
8	(8)	R	A (RAA)
8	(9)	$Q \vee R$	8,$\vee$I
3,8	(10)	$(Q \vee R) \wedge \neg(Q \vee R)$	3,9,$\wedge$I
3	(11)	$\neg R$	8,10,RAA
12	(12)	P	A (RAA)
1,12	(13)	Q	1,12,MPP
1,3,12	(14)	$Q \wedge \neg Q$	7,13,$\wedge$I
1,3	(15)	$\neg P$	12,14,RAA
1,2,3	(16)	R	2,15,MPP
1,2,3	(17)	$R \wedge \neg R$	11,16,$\wedge$I
1,2	(18)	$\neg\neg(Q \vee R)$	3,17,RAA
1,2	(19)	$Q \vee R$	18,DN

With TI the following shorter proof is also available:

1	(1)	$P \rightarrow Q$	A
2	(2)	$\neg P \rightarrow R$	A
	(3)	$P \vee \neg P$	TI 50
4	(4)	P	A ($\vee$E)
1,4	(5)	Q	1,4,MPP
1,4	(6)	$Q \vee R$	5,$\vee$I
7	(7)	$\neg P$	A ($\vee$E)
2,7	(8)	R	2,7,MPP
2,7	(9)	$Q \vee R$	8,$\vee$I
1,2	(10)	$Q \vee R$	3,4,6,7,9,$\vee$E.

Again, consider the use of TI in the proofs of

52   P ⊢ (P ∧ Q) ∨ (P ∧ ¬Q)

1	(1)	P	A
	(2)	Q ∨ ¬Q	TI(S) 50
3	(3)	Q	A(for ∨E)
1,3	(4)	P ∧ Q	1,3,∧I
1,3	(5)	(P ∧ Q) ∨ (P ∧ ¬Q)	4,∨I
6	(6)	¬Q	A(for ∨E)
1,6	(7)	P ∧ ¬Q	1,6,∧I
1,6	(8)	(P ∧ Q) ∨ (P ∧ ¬Q)	7,∨I
1	(9)	(P ∧ Q) ∨ (P ∧ ¬Q)	2,3,5,6,8,∨E

53   P → Q ⊢ P ∧ Q ↔ P

1	(1)	P → Q	A
	(2)	P ∧ Q → P	TI 47
3	(3)	P	A (CP)
1,3	(4)	Q	1,3,MPP
1,3	(5)	P ∧ Q	3,4,∧I
1	(6)	P → P ∧ Q	3,5,CP
1	(7)	(P → P ∧ Q) ∧ (P ∧ Q → P)	2,6,∧I
1	(8)	P ∧ Q ↔ P	7,df↔.

Taking sequents **26**

$$P ∧ (P ↔ Q) ⊢ P ∧ Q$$

and **52**

$$P ∧ Q ⊢ P ∧ (P ↔ Q)$$

we have

54   P → Q ⊣⊢ P ∧ Q ↔ P.

3.4.2 The Rule of Sequent Introduction (SI)

The Rule of Sequent Introduction will permit us to use any substitution instance of any sequent already proved as a *derived rule* to justify a line of a proof. To see that such a rule is admissable, assume that we have a proof of a sequent:

$$A_1, A_2, A_3 \vdash B.$$

Then by three applications of CP, we can produce a proof of the theorem:

$$A_1 \rightarrow (A_2 \rightarrow (A_3 \rightarrow B))$$

That theorem can now be cited in an application of TI or TI(S). For example, consider the proof of the sequent:

$$P \rightarrow (Q \rightarrow R), P, \neg R \vdash \neg Q$$

(which we will refer to as $$.)

1	(1)	$P \rightarrow (Q \rightarrow R)$	A
2	(2)	P	A
3	(3)	$\neg R$	A
4	(4)	Q	A (RAA)
1,2	(5)	$Q \rightarrow R$	1,2,MPP
1,2,4	(6)	R	4,5,MPP
1,2,3,4	(7)	$R \wedge \neg R$	3,6,$\wedge$I
1,2,3	(8)	$\neg Q$	4,7,RAA

Now to that proof add the lines:

1,2	(9)	$\neg R \rightarrow \neg Q$	3,8,CP
1	(10)	$P \rightarrow (\neg R \rightarrow \neg Q)$	2,9,CP
	(11)	$(P \rightarrow (Q \rightarrow R)) \rightarrow (P \rightarrow (\neg R \rightarrow \neg Q))$	1,10,CP

and call the theorem thus proved '¢¢'. Now suppose that we want a proof for the sequent:

$$S \rightarrow (T \rightarrow R), S, \neg R \vdash \neg T.$$

we could now make use of the newly proved theorem ¢¢ by TI(S) in such a proof as the following:

1	(1) $S \rightarrow (T \rightarrow R)$	A
2	(2) S	A
	(4) $(S \rightarrow (T \rightarrow R)) \rightarrow (S \rightarrow (\neg R \rightarrow \neg T))$	TI(S) ¢¢
1	(5) $S \rightarrow (\neg R \rightarrow \neg T)$	1,4,MPP
1,2	(6) $\neg R \rightarrow \neg T$	2,5,MPP
1,2,3	(7) $\neg T$	3,6,MPP

Thus, we may assert that

(a)   Every provable sequent has a corresponding theorem.
(b)   Any substitution instance of that theorem may be introduced.

Clearly every substitution instance of any provable sequent could be given a proof using TI or TI(S) in this way. Accordingly we abbreviate the procedure by leaving out what we know we can prove mechanically and proceed as follows:

1	(1) $S \rightarrow (T \rightarrow R)$	A
2	(2) S	A
3	(3) $\neg R$	A
1,2,3	(4) $\neg T$	1,2,3,SI(S) $$

In other words, we cite the proved sequent as though it were a rule. The justification mentions the input lines for the application, those that are substitution-instances of the premises of the previously proved sequent; then in the rule position, it gives SI or SI(S) together with an indication of where the cited sequent is proved. When the cited sequent is already known by some standard tag, such as 'DeMorgan's Theorem', an abbreviation of the tag (in this case 'DeM') can be used in the rule position of the justification without a sequent number. On the extreme left, the required assumptions are the required assumptions of the input lines. Sequents proved in previous exercises can also be cited using SI or SI(S).

In any system in which the set of provable sequents is closed under uniform substitution, the use of TI and SI greatly assists in the exploration of the system's properties, as theorems and sequents proved using TI and SI can themselves be cited in proofs of further theorems and sequents. We illustrate this fact below, when we have set out for easy reference all of the numbered sequents and theorems now available for use with TI and SI. The list excludes theorems and sequents proved as part of exercises, but these too may be cited in this role. Of course it is a matter of professional pride that one wants actually to be able to prove the theorems and sequents introduced in this way.

3.4.3 Sequents and Theorems Proved So Far

1   $P, P \rightarrow Q \vdash Q$

2   $P \rightarrow Q, Q \rightarrow R, P \vdash R$

3   $P \rightarrow (Q \rightarrow R), P, Q \vdash R$

4   $\neg\neg P \vdash P$

5   $P \rightarrow (Q \rightarrow R) \vdash Q \rightarrow (P \rightarrow R)$

6   $P, Q \vdash P \wedge Q$ (Adjunction)

7   $(P \wedge Q) \rightarrow R \vdash P \rightarrow (Q \rightarrow R)$ (Exportation)

8   $\vdash (P \wedge Q) \rightarrow P$ (Simplification)

9   $\vdash (P \wedge Q) \rightarrow Q$ (Simplification)

10   $P \rightarrow (Q \rightarrow R) \vdash (P \wedge Q) \rightarrow R$ (Importation)

11   $P \wedge Q \vdash Q \wedge P$ ($\wedge$-Commutation)

12   $Q \rightarrow R \vdash (P \wedge Q) \rightarrow (P \wedge R)$

13   $R, R \rightarrow Q \vdash P \rightarrow Q$

14   $P \vdash P \vee Q$

15   $P \vdash Q \vee P$

16   $P, (P \vee Q) \rightarrow R \vdash R$

17   $P \vee Q \vdash Q \vee P$ ($\vee$-Commutation)

18   $Q \rightarrow R \vdash (P \vee Q) \rightarrow (P \vee R)$

19   $P \vee (Q \vee R) \vdash Q \vee (P \vee R)$
20   $P \vee P \vdash P$
21   $P \rightarrow Q, P \rightarrow \neg Q \vdash \neg P$
22   $P \rightarrow \neg P \vdash \neg P$
23   $P \leftrightarrow Q \vdash Q \leftrightarrow P$ ($\leftrightarrow$-Commutation)
24   $P, P \leftrightarrow Q \vdash Q$
25   $P \leftrightarrow Q, Q \leftrightarrow R \vdash P \leftrightarrow R$ $P \rightarrow Q, Q \rightarrow R, P \rightarrow R$
26   $(P \wedge Q) \leftrightarrow P \vdash P \rightarrow Q$
27   $P \wedge (P \leftrightarrow Q) \vdash P \wedge Q$
28   $P \vdash P$
29   $(P \wedge Q) \rightarrow R \dashv\vdash P \rightarrow (Q \rightarrow R)$
30   $P \wedge (P \vee Q) \dashv\vdash P$
31   $P \vee (P \wedge Q) \dashv\vdash P$
32   $P \rightarrow Q, \neg Q \vdash \neg P$ (MTT)
33   $P \rightarrow \neg Q, Q \vdash \neg P$
34   $\neg P \rightarrow Q, \neg Q \vdash P$
35   $P \rightarrow Q, Q \rightarrow R, \neg R \vdash \neg P$
36   $P \rightarrow (Q \rightarrow R), P, \neg R \vdash \neg Q$
37   $P \rightarrow Q \vdash \neg Q \rightarrow \neg P$
38   $Q \rightarrow R \vdash (\neg Q \rightarrow \neg P) \rightarrow (P \rightarrow R)$
39   $P, \neg (P \wedge Q) \vdash \neg Q$ (MPT)
40   $P \rightarrow Q \dashv\vdash \neg (P \wedge \neg Q)$ $15, 1 (s) 40$
41   $P \vee Q \dashv\vdash \neg (\neg P \wedge \neg Q)$
42   $\neg (P \wedge Q) \dashv\vdash \neg P \vee \neg Q$ (DeM)
43   $\neg (P \vee Q) \dashv\vdash \neg P \wedge \neg Q$ (DeM)
44   $\vdash \neg (P \wedge \neg P)$ (Non-contradiction)
45   $\vdash P \rightarrow P$ (Identity Id)
46   $\vdash P \rightarrow \neg\neg P$
47   $\vdash \neg\neg P \rightarrow P$
48   $\vdash P \wedge Q \rightarrow P$
49   $\vdash (P \rightarrow Q) \rightarrow (\neg Q \rightarrow \neg P)$
50   $\vdash (P \rightarrow (Q \rightarrow R)) \rightarrow ((P \rightarrow Q) \rightarrow (P \rightarrow R))$
51   $\vdash P \vee \neg P$ (Excluded Middle (LEM))

52 $P \vdash (P \wedge Q) \vee (P \wedge \neg Q)$

53 $P \rightarrow Q \vdash P \wedge Q \leftrightarrow P$

54 $P \rightarrow Q \dashv\vdash P \wedge Q \leftrightarrow P.$

3.4.4 Some additional sequents and selected proofs

55 $\neg P \vee Q \vdash P \rightarrow Q$

1	(1)	$\neg P \vee Q$	A
1	(2)	$\neg(\neg\neg P \wedge \neg Q)$	1,SI(S) 40 ($\vdash$)
1	(3)	$\neg\neg P \rightarrow Q$	2,SI(S) 40 ($\dashv$)
4	(4)	P	A (CP)
4	(5)	$\neg\neg P$	4,DN
1,4	(6)	Q	3,5,MPP
1	(7)	$P \rightarrow Q$	4,6,CP

Taken together with the sequent proved in Exercise 2.6 (g), this gives us

56 $\neg P \vee Q \dashv\vdash P \rightarrow Q$ (Material Implication (MI)).

57 $P \vdash Q \rightarrow P$

1	(1)	P	A
1	(2)	$\neg Q \vee P$	1,$\vee$I
1	(3)	$Q \rightarrow P$	2,SI(S) 54

58 $\neg P \vdash P \rightarrow Q$

1	(1)	$\neg P$	A
1	(2)	$\neg P \vee Q$	1,$\vee$I
1	(3)	$P \rightarrow Q$	2,SI 54

In the previous two examples a citation of MI instead of SI, (SI(S)) **54** would have been adequate.

59 $\neg P, P \vee Q \vdash Q$ (Modus Tollendo Ponens (MTP))

1	(1)	$\neg P$	A
2	(2)	$P \vee Q$	A

3	(3)	P	A (∨E)
1	(4)	P → Q	1,SI 57
1,3	(5)	Q	3,4,MPP
6	(6)	Q	A (∨E)
1,2	(7)	Q	2,3,5,6,6,∨E

60   ¬Q, P ∨ Q ⊢ P (Modus Tollendo Ponens (MTP))

61   ⊢ (P → Q) ∨ (Q → P) (Connectedness (Con))

	(1)	P ∨ ¬P	LEM
2	(2)	P	A (∨E)
2	(3)	Q → P	2, SI 56
2	(4)	(P → Q) ∨ (Q → P)	3,∨I
5	(5)	¬P	A (∨E)
5	(6)	P → Q	5, SI 57
5	(7)	(P → Q) ∨ (Q → P)	6,∨I
	(8)	(P → Q) ∨ (Q → P)	1,2,4,5,7,∨E

The rules MPP, MTT taken together with the derived rules MPT and MTP are referred to as the *modi*. The last of these, MTP, is more commonly referred to nowadays as *Disjunctive Syllogism* (DS), 'syllogism' being an anglicized form of a word introduced by Aristotle as a label for what we would call an argument. Principles corresponding to all of the *modi* were in use at the time of the Stoics who seem to have taken them all as primitive. In contemporary logic, the principle MTP (or DS) is a subject of dispute in some branches of logical theory, although MPP is nevertheless accepted. Systems that accept the latter but lack the former do not admit the connection between '→' and '∨' expressed in the interderivability principle MI. In the system L, all are admitted.

The reader may earlier have wondered why, since it is provable, the ¬-introduction direction of DN was not admitted as a rule. We trust that by now the reason will have become clear. We want to adopt as *primitive* rules, only as many as we require. That a rule is provable as a sequent is sufficient reason *not* to include it in the initial set. However, since we now have the means to use proved sequents in effect as derived rules, there is no reason not to permit the use of the label, 'DN' for both the ¬-

elimination and the $\neg$-introduction directions. **Simon** provides switches under **Options, Rules** that will permit the $\neg$-introduction direction of DN to be cited as DN, and for citations of MTT as a rule. However, bear in mind that **Simon Says** is sensitive to these citations. They will be accepted only if they have been explicitly authorized by your course instructor.

Exercise 3.1

Use **Simon**'s proof editor to construct a proof for each of the following sequents, citing only the 9 primitive rules.

(a) $\vdash (Q \rightarrow R) \rightarrow ((P \rightarrow Q) \rightarrow (P \rightarrow R)$
(b) $\vdash P \rightarrow (Q \rightarrow P \wedge Q)$
(c) $\vdash (P \rightarrow R) \rightarrow ((Q \rightarrow R) \rightarrow (P \vee Q \rightarrow R))$
(d) $\vdash (P \rightarrow Q \wedge \neg Q) \rightarrow \neg P$
(e) $\vdash (\neg P \rightarrow P) \rightarrow P$

Exercise 3.2

Each of the following sequents is a substitution-instance of a sequent already proved. For each, cite by number the proved sequent of which it is a substitution-instance, and say what substitutions have been made.

(a) $(P \rightarrow Q) \rightarrow P, P \rightarrow Q \vdash P$
(b) $\neg\neg P \rightarrow \neg\neg\neg P, \neg\neg\neg\neg P \vdash \neg P$
(c) $\neg P \wedge (Q \wedge R) \rightarrow Q \vee P \vdash \neg P \rightarrow (Q \wedge R \rightarrow Q \vee P)$
(d) $(\neg Q \rightarrow Q) \rightarrow \neg(\neg Q \rightarrow Q) \vdash \neg(\neg Q \rightarrow Q)$
(e) $\neg(S \vee P) \vdash S \vee P \rightarrow (P \wedge Q) \leftrightarrow (R \vee S)$

Exercise 3.3

Use **Simon**'s proof editor to construct a proof for each of the following sequents, citing only the 9 primitive rules.

(a) $\vdash P \rightarrow (P \vee Q)$

Using this result, the primitive rules and TI, construct a proof for

(b) $Q \to P \vdash P \lor Q \leftrightarrow P$.

Taken together with Exercise 2.4(f) we now have

(c) $(P \lor Q) \leftrightarrow P \dashv\vdash Q \to P$.

Exercise 3.4

Use **Simon**'s proof editor to prove the following, again citing only primitive rules together with SI and **56**.

(a) $P \land Q \dashv\vdash P \land (P \leftrightarrow Q)$

Exercise 3.5

Use **Simon**'s proof editor to prove each of the following using primitive or derived rules together with any sequents or theorems already proved. Bear in mind that SI or SI(S) and TI or TI(S) can be used with citations of sequents and theorems proved in previous exercises.

(a) $\vdash P \lor (P \to Q)$
(b) $\vdash (P \to Q) \lor (Q \to R)$
(c) $\vdash ((P \to Q) \to P) \to P$
(d) $\neg Q \vdash P \to (Q \to R)$
(e) $P, \neg P \vdash Q$
(f) $P \lor Q \dashv\vdash \neg P \to Q$
(g) $\neg(P \to Q) \dashv\vdash P \land \neg Q$
(h) $(P \to Q) \to Q \vdash P \lor Q$
(i) $(P \to Q) \lor (P \to R) \dashv\vdash P \to Q \lor R$
(j) $P \to Q \dashv\vdash (P \leftrightarrow Q) \lor Q$
(k) $Q \vdash P \land Q \leftrightarrow P$
(l) $\neg Q \vdash P \lor Q \leftrightarrow P$

Exercise 3.6

Let A and B be any wffs. Demonstrate that $A \vdash B$ iff $\vdash A \to B$. Prove that $A \dashv\vdash B$ iff $\vdash A \leftrightarrow B$.

3.5 Effectiveness

Recall that we had set two goals: given an argument expressible in our language, we wanted to be able to find out by some finite procedure that it was a good argument if it was good (the positive requirement); we wanted to be able to find out by some finite procedure that it was bad if it was bad (the negative requirement). Both of these ambitions remain so far unfulfilled. We have given ourselves a system of rules by which to attempt to prove the conclusions of arguments from their premises. This was intended as a contribution to satisfying the positive requirement. But we have so far no assurance either that for every good argument there is such a proof, or that if there is, we could find it by a finite search. To know that there is such a proof would be helpful, or at least vaguely reassuring, but the knowledge would be of no practical use if we had no reassurance that we could actually find the proof. After all, we might try and try, but our failure to find a proof might have its source in our own proof-theoretic ineptness.

Now eventually we will find a practical resolution of the matter, but at the moment we give ourselves some purely technical reassurance. We can demonstrate that if an argument has a proof, we can find the proof by a finite (if somewhat impractical) procedure. The procedure depends upon the facts (a) that every wff is a finite string, and (b) that every proof is a finite sequence of wffs. Briefly, the procedure is this: we find a way of associating every proof uniquely and determinately with a natural number (one of 0, 1, 2, 3, . . .). Since every natural number is only finitely far along in the sequence of natural numbers, we would then know that every proof can (in principle) be discovered by a finite search of the natural numbers.

3.5.1 Arithmetizing the syntax

We achieve this association by translating our language into the language of arithmetic in such a way that the structures of sentences and derivatively, the structures of proofs are preserved. Since the translation preserves structure, it is referred to as an *arithmetization of the syntax* of our language. By this translation, each proof will be represented by the arithmetical name of a natural number from which it can be reconstituted in the language of L. We begin by translating each symbol of our language into a number, then translating wffs by concatenating (that is, writing without spaces) the translations of their symbols. Then we translate proofs by concatenating the translations of their wffs. Thus each proof will have a name in the language of arithmetic, a name which is also the name of a natural number. We would then (theoretically) search the proofs in the natural order of their arithmetic names. It is perhaps enough for our purposes to be told that such a procedure is possible, but, having been told that it is, there will naturally be an interest in seeing sufficiently many details to be able to figure out how it would be done.

3.5.2 Translating the symbols of *L*

We describe a translation that encodes each symbol of k and each propositional variable as a power of 10. Let τ (Greek TAU) be the translation that assigns to each symbol a power of $10 \leq 11$ as follows:

$$\tau(\neg) \qquad = 10^{10} \qquad = 10000000000$$
$$\tau(\rightarrow) \qquad = 10^{11} \qquad = 100000000000$$
$$\tau(\wedge) \qquad = 10^{12} \qquad = 1000000000000$$
$$\tau(\vee) \qquad = 10^{13} \qquad = 10000000000000$$
$$\tau(() \qquad = 10^{14} \qquad = 100000000000000$$
$$\tau()) \qquad = 10^{15} \qquad = 1000000000000000$$
$$\tau(P) \qquad = 10^{16} \qquad = 10000000000000000$$
$$\tau(P_1) \qquad = 10^{17} \qquad = 100000000000000000$$

$$\cdot$$
$$\cdot$$
$$\tau\{P_i\} \qquad = 10^{(i+16)}$$
$$\cdot$$

Now the number of each wff of Φ is the number whose numeral is the concatenation of the numerals of the numbers assigned to the individual symbols that make up the wff, in the order in which the symbols appear in it. Since no two distinct wffs have the same symbols in the same order, each wff is thus associated uniquely with a natural number. The order of the wffs in the enumeration is the natural order of the numbers thus associated with them.

Notice that in this assignment, we have restricted ourselves to atomic wffs consisting of 'P' with or without subscripts. Clearly this gives us a denumerably infinite set of atoms, as our description of *L* required. We can now compute $\tau(\alpha)$ for any wff α. Some examples will illustrate:

$\tau(\neg P) =$ 10   16
100000000001000000000000000000

$\tau((P \rightarrow P)) =$

1000000000000010000000000000000001000000000001000000000000000001000000000000000

14   16   11   16   15

and so on. Notice that since each symbol is translated into a sequence having a unique 1 in initial position, the original wff is unambiguously retrievable from its arithmetization.

Now let the rules, in some determinate order, be given as names, the names of the powers of 10 between 0 and 8 inclusively. With these resources, we can in principle give every proof a name that is also the name of a natural number. For each line of the proof, we (simply!) write the arithmetic names of its assumptions, followed immediately by the arithmetic name of its entry followed immediately by the arithmetic names of its inputs, followed by the arithmetic name of its rule. The arithmetic name of the proof is the concatenation of the arithmetic names of its lines. The order of the set of all proofs is the natural order of their arithmetic names. And again, as every proof has as its arithmetic name the name of some natural number, and since every such natural number is only finitely far along in the natural ordering of arithmetic names of proofs, every proof can be recovered after a finite search.

3.6 Validity

Well that, for an answer, is (as the poet[3] has said) what we might rightly call moderately satisfactory only. In the first place, it is at least tedious as a way of spending our time. In the second, we still do not know whether every good argument has a proof. In the third, we do not know whether

(

---

3. Henry Reed 'Lessons of the War (II. Judging Distances)' in *A Map of Verona*.

false positive

our system will let us construct proofs *only* for good arguments. We need some independent account of the goodness of arguments by which to judge the system itself on these last two counts. We want to be assured:

1. that all of the arguments for which the system will let us construct proofs are good arguments, and
2. that the system will let us construct a proof for any good argument.

Both require us to give some independent account of the *goodness* of an argument.

The classical logical answer to the question as to what makes an argument a good one is this: an argument is a good argument if and only if it is *valid*. An argument is *valid* if and only if all of its premisses cannot be true without its conclusion's being true as well. That is essentially the general account of validity we shall adopt here, but before we can say in detail how it applies to our language, we must say something about the conditions for truth and falsity of the wffs of the language of L. Bear in mind that until this paragraph the subject of truth has not yet arisen in this text, and in fact, our official account of the language speaks of wffs, that is, strings of symbols, not of sentences. Before we can speak of truth and falsity we must say something about how these strings are to be interpreted.

3.6.1 The interpretation

We have been referring to the atoms of our language as *propositional variables*. Historically they have been called *propositional* variables, because they have been supposed to vary over propositions, but no one has yet succeeded in saying what a proposition is. In our official interpretation, the atoms are variables that range over the set of values $\{0, 1\}$, which it will do no harm to think of the set of *truth-values*, that is as the set of values representing falsity (0) and truth (1). An assignment of 1 or 0 to every atomic wff can be thought of as a truth-value assignment. Since a wff has occurrences of only finitely many atoms, a single wff will permit us to distinguish only finitely many distinct classes of truth-value assignments. For example, if the only atoms occurring in a wff A are P and Q, then A will let us distinguish those assignments that assign 1 to both P and Q, those that assign 1 to P and 0 to Q, those that assign 0 to P and 1 to Q, and those that assign 0 to both. If for each of the connectives we can give a complete account of the conditions in which a wff formed with that connective takes 1, and if every such wff takes exactly one of the values in {1,0}, then we can give a complete account of the truth-conditions of every wff of the language of L. Thus every assignment of a truth-value to every propositional variable determines a truth-value for every sentence of the language. That is in fact how we proceed. We interpret each of the primitive connectives as an n-ary function that takes an n-tuple of truth-values as input and gives a truth-value as output, and we do this in such a way as to confirm the intuitive interpretations of the connectives that give us their readings. For example, ¬ is interpreted as a unary (or one-place) function that, given the input value 1, gives the output value 0, and given input value 0, outputs the value 1. Because this function takes truth-values as input and gives truth-values as outputs, it is called a *truth-function*. Intuitively we think of the behaviour of the function as reflecting the fact that if a sentence is true, then its negation is false and if a sentence is false, then its negation is true. The graphs of such functions are tabulated in tables called *truth-tables*. The truth-table for negation is:

$$¬A$$
$$0 \quad 1$$
$$1 \quad 0$$
$$↑$$

The input values are given under the initial A; the output value corresponding to each input value is given in the same row in the principle column, indicated by ↑, that is, the column under the main connective (the connective of longest scope).

The three primitive binary connectives are interpreted as *binary truth-functions*, that is, as functions taking pairs of values as inputs and giving truth-values as outputs. The tables displaying their graphs are constructed in the same manner, the input pair of values at each row is the pair of the values under the metalogical variables in that row. The corresponding output value is that under the main connective:

¬A ∨ B   = A → B   A ∧ B   A ∨ B
              1 1 1   1 1 1   1 1 1
              1 0 0   1 0 0   1 1 0
              0 1 1   0 0 1   0 1 1
              0 1 0   0 0 0   0 0 0
                 ↑   ↑   ↑

[handwritten left margin: Remember: A conditional is 0 only when B is 0 and A is 1.]

The table for ∧ reveals that a conjunction will take the value 1 (is true) when both of its conjuncts take the value 1 (are true); else the conjunction takes the value 0 (is false). From the table for ∨ we see that a disjunction takes the value 0 when both of its disjuncts take 0; otherwise the disjunction takes 1. We can sum up the entries in these tables equivalently by saying that ∧ outputs the lesser of its two input values; ∨ outputs the greater of its two input values. The table for → interprets that connective as the truth-function historically known as *the material conditional*. It outputs the value 0 when its antecedent has a 1 and its consequent has a 0; else it outputs a 1. This is sometimes expressed by saying that the material conditional preserves truth. The reading is evident enough for the sole case in which 0 is the output and for the case in which both antecedent and consequent are true; for the other two cases, those where the antecedent has a 0, it is said, somewhat colourfully, to preserve all of the truth that there is to preserve. Certainly in those cases, it cannot be said to fail to preserve truth, since there is no truth for it to fail to preserve. The tag *material* is intended to convey the feature of the conditional that its truth depends upon the material facts of the case, rather than upon some special connection between antecedent and consequent. In this regard, it may be helpful to attend to the connection between A → B and ¬A ∨ B, which we know from the proof of Sequent **55** to be deductively equivalent. Our interpretation of ∨ is such as to make ¬A ∨ B true exactly when at least one of ¬A, B is true. Thus is ¬A ∨ B is true, then *if ¬A is not true, then B is*. Note that the italicized portion of that sentence is in *if-then* form: it is also equivalent (by the table for ¬) to *if A is true, then B is true*. In fact the principle column of the table for → is exactly the principle column for the table for ¬A ∨ B, (whose input values are those under the ¬ and under B.)

Since all wffs are constructed using only those connectives, which we have now interpreted as truth-functions, every wff can now be interpreted as compositions of those fundamental functions. (The composition of two functions f and g is the function $g \circ f$, where $g \circ f(x)$ is $g(f(x))$, that is $g \circ f$ is the function that outputs the outputs of g for inputs that are the outputs of f.) Thus for example, the binary truth-function that interprets $\neg A \vee \neg B$ takes the pair of outputs of the unary $\neg$ truth-function for A and for B as the inputs to the binary $\vee$ truth-function and, as required, outputs the greater of those two values. The values in the principal column, in standard order, will be 0111. In general, the computation of output values for these composed functions takes output values for functions representing shortest-scope occurrences of connectives as input values of functions representing next-longer-scope occurrences of connectives and so on until the function corresponding to the longest-scope occurrence of a connective finally outputs a value for the wff as a whole. The number of distinct inputs to an n-ary truth function is 2^n: thus the table for an n-ary truth-function will have 2^n rows. So, for example, the ternary function that interprets the wff $(P \vee Q \rightarrow Q \wedge R)$ would have as its table:

$$
\begin{array}{ccccccc}
P & \vee & Q & \rightarrow & Q & \wedge & R \\
1 & 1 & 1 & 1 & 1 & 1 & 1 \\
1 & 1 & 1 & 0 & 1 & 0 & 0 \\
1 & 1 & 0 & 0 & 0 & 0 & 1 \\
1 & 1 & 0 & 0 & 0 & 0 & 0 \\
0 & 1 & 1 & 1 & 1 & 1 & 1 \\
0 & 1 & 1 & 0 & 1 & 0 & 0 \\
0 & 0 & 0 & 1 & 0 & 0 & 1 \\
0 & 0 & 0 & 1 & 0 & 0 & 0 \\
\end{array}
$$

there being 2^3 or 8 combinations of truth-values of its atoms understood as variables taking values from $\{0,1\}$. The truth-table for the interpretation of a wff having four distinct atoms would have sixteen rows, and so on.

To construct a truth-table for a wff, first enter all possible inputs to the function. To do this, count the number of propositional variables in the wff. If there are n distinct variables, then the table will have 2^n rows. The canonical order for filling in such a table is to fill the first half of the places ($2^n/2$ places) under the leftmost variable with 1's, and the second half of them with 0's. Then under the next variable alternate groups of $2^n/4$ 1's and $2^n/4$ 0's, and so on until under the last variable there are alternating 1's and 0's. Then fill in the output values under the connectives of shortest scope, according to their own tables, then take those values as inputs to the connectives of next longest scope, filling in the spaces under them according to their tables, and so on until the output values under the connective of longest scope are filled in. In the last table, which is for a truth-function in three variables, we find alternating groups of $2^3/2 = 4$ 1's and 0's under the P, groups of $2^3/4 = 2$ 1's and 0's under the Q, and then alternating 1's and 0's under the R. We then fill in the output values under the ∨ taking the pairs of values under P and Q as inputs. Here we enter the greater of each pair of input values. Then we fill in the output values under the ∧ taking the pairs of values under Q and R as inputs. Here we enter the smaller of each pair of input values. Finally, we enter the values under the →: a 0 when P ∧ Q has the value 1, and Q ∨ R has the value 0, a 1 in all other cases.

Note on Simon: **Simon** includes a truth-table editor that automates much of the work of constructing such tables.

3.6.2 Constant truth-functions

Notice that the principal column of the table for the interpretation of (P ∨ Q → Q ∧ R) has some 0's and some 1's. This function outputs 1 for the input triple of values <1,1,1> but outputs 0 for the input <1,1,0>. Such a truth-function, one that outputs both 1's and 0's, is said to be *contingent*, the label indicating that its outputs are contingent upon its inputs. Contrast that case with the table for interpretation of the wff (P ∧ Q → P ∨ Q).

$$P \land Q \rightarrow P \lor Q$$
$$1\ 1\ 1\quad 1\ 1\ 1\ 1$$
$$1\ 0\ 0\quad 1\ 1\ 1\ 0$$
$$0\ 0\ 1\quad 1\ 0\ 1\ 1$$
$$0\ 0\ 0\quad 1\ 0\ 0\ 0$$
$$\uparrow$$

The truth-function that interprets the wff (P $\land$ Q $\rightarrow$ P $\lor$ Q) is a *constant truth-function*; that is, its output is the same for every input. In the case of this wff, the output is constantly 1; that is, its output is 1 for all inputs. A constant truth-function, the output value of which is constantly 1, is said to be a *tautologous* or *tautological* truth-function. We may contrast that case in turn with the truth-function that interprets the negation of (P $\land$ Q $\rightarrow$ P $\lor$ Q), the output of which must accordingly be constantly 0. Such a truth-function is said to be an *unsatisfiable* truth-function. In both cases, the label given to the truth-function that interprets the wff is also applied to the wff itself. (P $\land$ Q $\rightarrow$ P $\lor$ Q) is said to be a *tautology* or a *tautologous wff* ; $\neg$(P $\land$ Q $\rightarrow$ P $\lor$ Q) is said to be an *unsatisfiable wff*.

To test a wff for tautologousness, we do not need to construct the whole table of its truth-functional interpretation. We need only check to see whether there could be a 0 among its output values, that is, in the principal column of its table. This point is of some practical value, as a wff having occurrences of 32 distinct atoms will be interpreted as a truth-function having a table with approximately 4.29 billion rows. As a manageable   example,   we   can   consider   the   wff (P $\land$ Q $\rightarrow$ R) $\rightarrow$ (P $\rightarrow$ R) $\lor$ (Q $\rightarrow$ R). If we first place a 0 in its principal column, and then obey the dictates of that entry, we will produce the following

$$(P \land Q \rightarrow R) \rightarrow (P \rightarrow R) \lor (Q \rightarrow R).$$
$$1\ 0\ 0\quad 1\ 0\quad\ 0\ 1\quad 0\ 0\quad 0\ 1\quad 0\ 0$$
$$7\ 6\ 8\quad 2\ 5\quad\ 1\ 4\quad 3\ 4\quad 2\ 4\ 3\ 4$$
$$\ \ \ \ *\quad\quad\quad\quad\ \ \uparrow\quad\quad\quad\quad\quad *$$

The numerals at the bottom have been included here only to indicate the order of the entries of 1's and 0's in the purported row above, and would ordinarily not appear. The *'s indicate that the initial entry of the 0 in the main column (marked with ↑) has dictated an entry of 1 under one occurrence of Q and an entry of 0 under another entry of Q. That is, the supposition that the output value of the truth-function is 0 has led us to the conclusion that a propositional variable has simultaneously taken two distinct values. Since this conclusion is absurd, the supposition that led to it is false. Therefore the truth-function cannot have the output value 0; therefore its output value is constantly 1; therefore the wff

$$(P \wedge Q \to R) \to (P \to R) \vee (Q \to R)$$

is a tautology.

Now an initial entry of 0 as the output of a truth-function may not dictate any particular entry at any particular stage. For example, an entry of 0 under the main connective of

$$(P \vee Q \to R) \to (P \to R) \wedge (Q \to R)$$

would dictate a 0 under the ∧ of the consequent, but would not dictate any particular value under the conjuncts, only that a 0 must occur under one of them. In this case, where either of two entries would be permitted, *both possibilities must be checked.* If each can be shown to require that some propositional variable should simultaneously take both the value 1 and the value 0, we may conclude that it is impossible for that truth-function to output a 0. In the case in question, an entry of 0 under either conjunct requires that R have the value 0 and that either P or Q have the value 1. But a 1 under either P or Q will give a 1 under the ∨ in the antecedent. But that 1 together with the 1 under the → (required by the initial 0) forces a 1 under the occurrence of R in the antecedent. Thus the two occurrences of R have different values. Hence the initial 0 represents an impossible output, whence it follows that the wff is a tautology. If the initial entry of 0 leads to no such conflicting entries, this indicates that there is an input for which the function gives 0 as output, and the wff is therefore not a tautology.

By corresponding methods we can test whether a wff is unsatisfiable: Assume that there is a 1 in the output of the function and follow the dictates of that assumption by making entries in a suppositious row of its table. If the entry leads to conflicting entries under two occurrences of the same atom, then there can be no 1 in the output of the function. Therefore the output is constantly 0, whence the wff is unsatisfiable. Consider the following example:

$$\neg(P \rightarrow Q) \wedge \neg(Q \rightarrow P)$$

A 1 under the '$\wedge$' will require 1's under both occurrences of '$\neg$'. This will in turn require 0's under both occurrences of '$\rightarrow$'. But the first of those 0's will require a 1 under the first occurrence of 'P', the second a 0 under the second occurrence of 'P'. Therefore the initial entry of 1, representing an output value of 1 for the truth-function as a whole requires that the atom P simultaneously take both the value 1 and the value 0. Therefore the truth-function cannot output a 1; therefore it constantly outputs 0. Therefore it is unsatisfiable.

However, the initial entry may simply lead to the discovery of an input for which 1 is the output, in which case the wff is not unsatisfiable. Such would be the case for the wff

$$(P \rightarrow Q) \wedge (Q \rightarrow P).$$

Note on Simon: **Simon** includes an assignment search editor that automates much of the work of searching for such assignments. When an attempted assignment forces an inconsistent assignment to an atom, the conflict is indicated by a rapid alteration of truth-values under that atom.

3.6.3 Substitution-instances

Since a tautologous truth-function outputs 1 for every input, it will be obvious that if A is a tautologous wff, then any substitution-instance of A is also a tautologous wff. Recall that a substitution-instance of a wff is a wff resulting from some succession of uniform substitutions of wffs for atoms. Clearly no such substitution can result in an input set of values not already represented in the truth-table of the original formula. Hence no new output values can result from any such substitution.

The corresponding observation can be made for unsatisfiable wffs. Since the truth-function representing it outputs 0 for all input sets of values, and no new input sets of values can result from the substitution, the output of the resulting function will be constantly 0.

The observation notably does not apply to contingent truth-functions, since substitutions may restrict the input sets of values to those for which the output is 1 or to those for which the output is 0. As an obvious example, consider the contingent truth-function representing P ∨ Q, which outputs 1 for three input pairs of values, and outputs 0 for the pair <0,0>. Substituting ¬P for the only occurrence of Q results in the tautological truth-function P ∨ ¬P. For this function, the input pair <0,0> cannot occur. In general, a contingent wff will have both tautologous and unsatisfiable substitution-instances. If the wff is contingent, then it has both output values. Consider a row in its table for which it gives the output value 1. Now for every atom that has the value 1 in that row, substitute a tautology for every occurrence of that atom; for every atom that has the value 0 in that row, uniformly substitute an unsatisfiable wff. The resulting substitution-instance must always give the output value 1. Now consider a row in its table for which it gives the output value 0. For every atom that has the value 1 in that row, substitute a tautology for every occurrence of that atom; for every atom that has the value 0 in that row, uniformly substitute an satisfiable wff. The resulting substitution-instance must always give the output value 0.

3.6.4 The number of n-ary truth-functions

We close this initial discussion of interpretation with some general remarks about n-ary truth-functions. We have already remarked that the tabulation of an n-ary truth-function will have 2^n rows. This fact also gives us the answer to the question: how many n-ary truth-functions are there? The answer is that there are as many n-ary truth-functions as there are distinct ways of filling up a column having 2^n positions. Since there are 2 ways of filling in each position, there are 2^{2^n} distinct ways of filling up the column, and hence that many n-ary truth-functions. Thus for example, there are four unary truth-functions

A	α	β	γ	δ
1	1	1	0	0
0	1	0	1	0

Of these, α is A ∨ ¬A; β is A; γ is ¬A, δ is A ∧ ¬A. The first and the last are constant functions: the former tautologous, the latter unsatisfiable. The other two are contingent.

There are sixteen binary truth-functions, represented by a – p in the following table:

A B	a	b	c	d	e	f	g	h	i	j	k	l	m	n	o	p
1 1	1	1	1	1	1	1	1	1	0	0	0	0	0	0	0	0
1 0	1	1	1	1	0	0	0	0	1	1	1	1	0	0	0	0
0 1	1	1	0	0	1	1	0	0	1	1	0	0	1	1	0	0
0 0	1	0	1	0	1	0	1	0	1	0	1	0	1	0	1	0

Some of these will be familiar from our discussion so far: b represents ∨; e represents →; h represents ∧. Derivatively, g is recognizable as the representative of ↔, given the representatives of → and ∧. In fact, as we shall shortly discover for ourselves, the language that we have adopted is expressively sufficient for all such truth-functions. That is, each of the sixteen functions represents a class of wffs of the language of L. For example, j represents the composition ¬∘↔. That is, A j B expresses ¬(A ↔ B), and so on.

Exercise 3.7

Use **Simon**'s truth-table editor to construct a truth-table for each of the following wffs.

(a) P ∧ Q → Q ∧ P
(b) Q → R → ((P ∧ Q) → (P ∧ R))
(c) (P → Q) ∧ (R → S) → (P ∨ S → Q ∨ R)

Exercise 3.8

For each of the following wffs, use **Simon**'s assignment search
editor to test it for tautologousness, unsatisfiability, or
contingency. In each case say which it is.

(a) $P \rightarrow P$

(b) $P \rightarrow \neg P$

(c) $\neg (P \rightarrow P)$

(d) P

(e) $\neg P \rightarrow (P \rightarrow Q)$

(f) $(P \leftrightarrow Q) \leftrightarrow \neg (P \leftrightarrow \neg Q)$

(g) $(P \wedge Q) \wedge \neg (P \leftrightarrow Q)$

(h) $(P \vee \neg Q) \wedge \neg (\neg P \rightarrow \neg Q)$

(i) $(P \rightarrow Q) \wedge (R \rightarrow S) \rightarrow (P \vee R \rightarrow Q \vee S)$

For each contingency discovered in the previous examples, give both a
tautologous substitution-instance and an unsatisfiable substitution-
instance.

Exercise 3.9

For each of the 16 binary truth-functions, find a wff having
occurrences of no connectives other than (a) $\neg$ and $\vee$, (b) $\neg$ and
$\wedge$, (c) $\neg$ and $\rightarrow$.

Exercise 3.10

Let | (the Sheffer Stroke function) be the binary truth-function i,
and let $\downarrow$ (the dagger stroke function) be the binary truth-
function o. For each of the 16 binary truth-functions show how it
can be expressed (a) using | alone (brackets permitted) (b) using
$\downarrow$ alone (again brackets permitted).

Exercise 3.11

Let A be a wff containing any number of atoms but with $\wedge$ as its
only connective. Show that A must be contingent.

Exercise 3.12

> Let A be a wff containing any number of atoms but with $\lor$ as its
> only connective. Show that A must be contingent.

3.7 The Strong Soundness of L

The interpretation of the language of L outlined in the last few sections
has provided what would be referred to as a *truth-theory* (as distinct from
the *proof-theory*) for the system L. With that truth-theory in hand, we are
in a position to make precise the informal notion of validity introduced
earlier (p. 72). Recall that that notion of validity was one according to
which an argument is valid iff its premises cannot be true without its
conclusion's also being true. We can now make precise the force of the
cannot of that formulation. For arguments expressible in the language of
L, to say that the premises cannot be true without the conclusion's being
true is to say that there is no assignment of truth-values to the atoms of
the argument that makes the premises true and the conclusion false. The
relationship that obtains between an ensemble of wffs and a single wff is
called *entailment*.

An ensemble Γ *entails* a wff A iff every assignment of truth-values to the
atoms of the language that makes all of the wffs of Γ true, makes A true.
We write $\Gamma \vDash A$ for 'Γ entails A.' Entailment is the truth-theoretic
counterpart of provability. In the particular case in which $\Gamma \vDash A$, and
$\Gamma = \varnothing$, we write $\vDash A$, and read this 'A is *valid* (on the interpretation).'
Thus the truth-theoretic counterpart of the proof-theoretic notion of
theoremhood is the notion of validity. In the proposed interpretation the
wff A is valid if and only if A is a tautology.

A system is said to be *strongly sound with respect to an interpretation*
iff, for every ensemble of wffs, Γ and every wff A, if $\Gamma \vdash A$, then $\Gamma \vDash A$
on that interpretation. It is said to be weakly sound with respect to the
interpretation if and only if every theorem is a valid wff. L is thus
weakly sound with respect to the proposed interpretation if and only if
every L-theorem is a tautology. This then is the precise formulation of
the independent standard to which a formal natural deductive system

must be held. Since the truth-functional interpretation outlined is the intended interpretation of the system L, we must show that L is sound with respect to that interpretation; we must demonstrate that for any ensemble Γ, and for any wff A, if there is a proof of A from Γ, then any assignment that makes all of the wffs of Γ true must also make A true. Our next step is to produce such a demonstration for L.

3.7.1 The Strategy of the demonstration

Recall the definition of an L-proof of A from an ensemble Γ. It is a finite sequence of wffs, $B_1, \ldots, B_i, \ldots, B_n$, where B_n is A, each B_i is justified by a rule of L, and the required assumptions for B_n are all wffs of Γ. Now if we can demonstrate that for any such proof, the required assumptions for B_n entail B_n, then our case is made. For if every assignment that makes all of those assumptions true also makes A true, then any assignment that makes all the wffs of Γ true also makes A true. That then will be our strategy. We demonstrate that any assignment that makes all of the required assumptions at the last line of a proof true must also make the last wff of the proof true.

That is the strategy. We also need tactics. These will be the following: we will demonstrate that *the property holds for every line of a proof.* That is, we will show that for every line of a proof, the ensemble of the required assumptions of the line entails the wff entered at that line.

Those are the tactics. We also need a method. This will follow a procedure called *strong mathematical induction.* We will demonstrate two jointly sufficient results:

(a)   the property holds of the first line of a proof (the basis of the induction),

(b)   for any k, if the property holds for every line of the proof before line k, then it holds for line k as well (the inductive step).

From these two results the tactically desired result follows immediately: the property holds for every line of the proof. (Since it holds for line 1 (by (a)), it holds for all lines before line 2. Therefore it holds for line 2, and therefore for all lines before line 3, and so on.)

Since the property holds for every line of the proof, in particular, it holds for the last line, and this gives the desired result. An assignment that makes all of the premises of the provable sequent true also makes the conclusion true. That is, the premises entail the conclusion.

3.7.2 The proof in detail

For arbitrary ensemble Γ and arbitrary wff A, assume that $\Gamma \vdash A$. Let the proof of A be $B_1, \ldots B_k \ldots, B_n$. We prove by induction that for every i $(1 \leq i \leq n)$, the ensemble Γ_i of the required assumptions at line i entails B_i.

The Basis
Let i = 1. Then line i is justified by the Rule of Assumption. Then the sole member of Γ_i is B_1. Trivially, then $\Gamma_i \vDash B_i$ when i = 1.

The Inductive Step

We must prove that IF for every i < k, $\Gamma_i \vDash B_i$, THEN $\Gamma_k \vDash B_k$. We assume the IF-clause, and then demonstrate on this assumption that the THEN-clause is also true. The assumed if-clause is called the *Hypothesis of Induction (H.I.)*.

Hypothesis of Induction (H.I.): For every i < k, $\Gamma_i \vDash B_i$.

Let i = k. (We prove that $\Gamma_k \vDash B_k$.)
Now B_k is a line of a proof; therefore it is justified by one of the nine rules of L. We demonstrate for each of the rules, that if B_k is justified by that rule, $\Gamma_k \vDash B_k$ (given H.I.). We consider each rule in turn.

Case 1. B_k is justified by the Rule of Assumption. Then the argument is as for the basis.

Case 2. B_k is justified by Modus Ponendo Ponens. Then the proof must have the following lines:

$$\Gamma_i \qquad \text{(i)} \qquad B_i \qquad\qquad \text{<rule> etc.}$$

$$\Gamma_j \qquad \text{(j)} \qquad B_i \to B_k \qquad \text{<rule> etc.}$$

$$\Gamma_k \qquad \text{(k)} \qquad B_k \qquad\qquad \text{i,j,MPP}$$

Since B_k is justified by MPP, Γ_k is Γ_i, together with Γ_j. Assume an arbitrary assignment (of truth-values to atoms) that makes all of the assumptions of Γ_k true. Then that assignment makes all the assumptions of Γ_i true and all the assumptions of Γ_j true. But by H.I., $\Gamma_i \vDash B_i$, and $\Gamma_j \vDash B_i \to B_k$. But by the truth-table for $A \to B$, if the antecedent is true, and the conditional is true, then the consequent is true. Therefore, B_k is true. But the assignment was arbitrary; therefore any assignment of truth-values to atoms that makes all of the required assumptions of Γ_k true makes B_k true if B_k is justified by Modus Ponendo Ponens. That is, if B_k is justified by MPP, then $\Gamma_k \vDash B_k$.

Case 3. B_k is justified by The Rule of Double Negation. Then the proof must have the lines

$$\Gamma_j \qquad \text{(j)} \qquad B_j (= \neg\neg A) \qquad \text{<rule> etc.}$$

$$\Gamma_k \qquad \text{(k)} \qquad B_k (= A) \qquad \text{i,j,DN}$$

In this case, $\Gamma_k = \Gamma_j$. Assume an assignment that makes all the assumptions of Γ_k. Then that assignment makes all of the assumptions of Γ_j true. By H.I., that assignment makes B_j true. By the truth-table for $\neg A$, A is true if and only if $\neg\neg A$ is true. Therefore B_k is true. But the assignment was arbitrary; therefore any assignment that makes all of the assumptions of Γ_k true also makes B_k true if B_k is justified by DN. That is, if B_n is justified by DN, then $\Gamma_k \vDash B_k$.

Case 4. B_k is justified by The Rule of Conditional Proof. Then the proof must have the following lines:

Γ_i (i) B_i A

               .

Γ_j (j) B_j \<rule\> etc.

               .

Γ_k (k) $B_k (= (B_i \rightarrow B_j))$ i,j,CP

Since B_k is justified by CP, Γ_k is Γ_j with B_i removed, and B_i is the sole member of Γ_i, we must demonstrate that any assignment (of truth-values to atoms) that makes all of the assumptions of Γ_k true and also makes B_i (the antecedent of B_k) true, makes B_j (the consequent of B_k) true. Assume an arbitrary assignment that makes all of the assumptions of Γ_k true, and also makes B_i true. Then that assignment makes all the assumptions of Γ_i true and therefore all the assumptions of Γ_j true. But by H.I., $\Gamma_j \vDash B_j$. Therefore, B_j is true; therefore $B_k (= (B_i \rightarrow B_j))$ is true. But the assignment was arbitrary; therefore any assignment of truth-values to atoms that makes all of the required assumptions of Γ_k true makes B_k true if B_k is justified by The Rule of Conditional Proof. That is, if B_k is justified by CP, then $\Gamma_k \vDash B_k$.

Case 5. B_k is justified by The Rule of $\wedge$-Introduction. Then the proof must have the following lines:

Γ_i (i) B_i \<rule\> etc.

               .

Γ_j (j) B_j \<rule\> etc.

               .

Γ_k (k) $B_k (= (B_i \wedge B_j))$ i,j,$\wedge$I

Then Γ_k is Γ_i together with Γ_j. Now assume an assignment that makes all of the assumptions of Γ_k true. Then that assignment makes all of the assumptions of Γ_i true, and all the assumptions of Γ_j true. Then by H.I.

that assignment makes B_i true and B_j true. By the truth-table for $\land$, that assignment also makes B_k $(= (B_i \land B_j))$ true. But that assignment was arbitrary. Therefore, any assignment that makes all the assumptions of Γ_k true also makes B_k true if B_k is justified by $\land$-introduction. That is, if B_k is justified by $\land$I, then $\Gamma_k \vDash B_k$.

Case 6. B_k is justified by The Rule of $\land$-Elimination. Then the proof must either have the lines:

Γ_j (i) $B_j (= (A \land C))$ <rule> etc.

$\quad\quad$.

Γ_k (k) $B_k (= (A))$ j,$\land$E

or the lines

Γ_j (i) $B_j (= (A \land C))$ <rule> etc.

$\quad\quad$.

Γ_k (k) $B_k (= (C))$ j,$\land$E.

In this case Γ_k is identical to Γ_j. Assume an assignment that makes all the assumptions of Γ_k true. Then that assignment makes all the assumptions of Γ_j true. By H.I. that assignment also makes $(A \land C)$ true. Then by the truth-table for $\land$, this makes A true and also makes C true. Therefore that assignment also makes B_k true. But the assignment was arbitrary. Therefore any assignment that makes all the assumptions of Γ_k true makes B_k true if B_k is justified by the Rule of $\land$-Elimination. That is, if B_k is justified by $\land$E, then $\Gamma_k \vDash B_k$.

Case 7. B_k is justified by the Rule of $\lor$-Introduction. In this case the proof must have either the lines:

Γ_j (j) B_j <rule> etc.

$\quad\quad$.

Γ_k (k) $B_k (= (C \lor B_j))$ j,$\lor$I or the lines

Γ_j (j) B_j \<rule\> etc.

Γ_k (k) $B_k (= (B_j \vee C))$ j,$\vee$I.

In this case Γ_k is identical to Γ_j. Assume an assignment that makes all the assumptions of Γ_k true. Then that assignment makes all the assumptions of Γ_j true. By H.I. that assignment also makes B_j true. Then by the truth-table for $\vee$, this makes $C \vee B_j$ true and also makes $B_j \vee C$ true. Therefore that assignment also makes B_k true. But the assignment was arbitrary. Therefore any assignment that makes all the assumptions of Γ_k true makes B_k true if B_k is justified by the Rule of $\vee$-Introduction. That is, if B_k is justified by $\vee$I, then $\Gamma_k \models B_k$.

Case 8. B_k is justified by the Rule of $\vee$-Elimination. In this case the proof must have the lines:

Γ_e (e) $B_e (= (A \vee C)$ \<rule\> etc.

Γ_f (f) $B_f (= A)$ A

Γ_g (g) $B_g (= B_k)$ \<rule\> etc.

Γ_j (h) $B_j (= C)$ A

Γ_i (i) $B_i (= B_k)$ \<rule\> etc.

Γ_k (k) B_k e,f,g,h,i,$\vee$E.

In this case Γ_k is Γ_e together with Γ_g without B_f and Γ_i without B_j. Assume an arbitrary assignment that makes all of the assumptions of Γ_k true. Then that assignment makes all the assumptions of Γ_e true. By H.I., that assignment makes B_e (= (A ∨ C)) true. By the truth-table for ∨, either that assignment makes A true or that assignment makes C true. Assume that it makes A true. Then it makes all the assumptions of Γ_g true; therefore it makes B_g (= B_k) true. Assume that it makes C true. Then it makes all the assumptions of Γ_i true; therefore it makes B_i (= B_k) true. That is, in either case the assignment makes B_k true. But the assignment was arbitrary, therefore any assignment that makes all the assumptions of Γ_k true makes B_k true if B_k is justified by the Rule of ∨-Elimination. That is, if B_k is justified by ∨E, then $\Gamma_k \vDash B_k$.

Case 9. B_k is justified by Reductio Ad Absurdum. Then the proof must have the following lines:

$$\Gamma_i \quad \text{(i)} \quad B_i \qquad\qquad\qquad A$$

$$\cdot$$

$$\Gamma_j \quad \text{(j)} \quad B_j \,(= (A \wedge \neg A)) \qquad \text{<rule> etc.}$$

$$\cdot$$

$$\Gamma_k \quad \text{(k)} \quad B_k \,(= \neg B_i) \qquad\qquad \text{i,j,RAA.}$$

In this case Γ_k is Γ_j without B_i. Assume an assignment that makes all the assumptions of Γ_k true, but which does not make B_k true. Then, by the truth-table for ¬, that assignment makes B_i true. But then that assignment makes all of the assumptions of Γ_j true. Now by H.I., $\Gamma_j \vDash B_j$. Therefore that assignment makes B_j (= (A ∧ ¬A)) true, which is absurd. Therefore no assignment that makes all the assumptions of Γ_k true makes B_k false if B_k is justified by Reductio Ad Absurdum. That is, if B_k is justified by RAA, then $\Gamma_k \vDash B_k$.

Since B_k must be justified by one of the nine rules, and since in each case $\Gamma_k \vDash B_k$ if for every i < k, $\Gamma_i \vDash B_i$, we may conclude that the inductive step has been demonstrated: if for every i < k, $\Gamma_i \vDash B_i$, then $\Gamma_k \vDash B_k$. This completes the induction. We therefore conclude that for every i ($1 \leq i \leq$ n), $\Gamma_i \vDash B_i$. In particular $\Gamma_n \vDash B_n$. Therefore $\Gamma \vDash A$.

But Γ and A were arbitrary, therefore for every ensemble Γ of wffs, and every wff A, if $\Gamma \vdash A$, then $\Gamma \vDash A$. That is,

Metatheorem 3.2

L is strongly sound with respect to its standard truth-functional interpretation.

3.7.3 Consequences of strong soundness

One immediate corollary of strong soundness is weak soundness. Since the most recent metatheorem was proved for any ensemble of wffs and any wff, we may infer that it holds in the special case in which Γ is $\varnothing$, the empty ensemble. We may therefore assert:

Corollary 3.0.1

L is weakly sound with respect to the interpretation given.

That is, for every wff A, if $\vdash A$, then $\vDash A$. That is, every L-theorem is a tautology.
Since no contradiction can be made true by any assignment of truth-values to atoms, we can also infer:

Corollary 3.0.2

The system L is consistent.

That is, no contradiction is an L-theorem.

Now the Strong Soundness Metatheorem also furnishes us with the finite negative procedure that was one of the historical goals of logical theory. If an argument, expressible in the language of L is a bad argument, then we have a finite procedure for discovering that it is bad. If the argument has occurrences of n distinct atoms, then a search of 2^n assignments of truth-values to atoms will exhibit the badness of the argument if it is bad, for if it is bad, then at least one assignment of truth-values to atoms will make its premises true and its conclusion false. Furthermore, by the Strong Soundness Metatheorem, we can conclude that the argument has no proof in L.

Suppose, by contrast, that an examination of those 2^n assignments of truth-values to atoms reveals that the ensemble of the premises of the argument entails the conclusion. Can we infer that the system L will provide a proof of the conclusion from the ensemble of the premises? That question is the subject of the next section.

Exercise 3.13

For each of the following sequents demonstrate that it has no proof in the system L.

(a) $P \wedge Q \to R \vdash P \to R$
(b) $P \to Q \vee R \vdash P \to Q$
(c) $P \to Q, P \to R \vdash Q \to R$
(d) $P \to R, Q \to R \vdash P \to Q$
(e) $P \to (Q \to R), Q, \neg R \vdash P$
(f) $P \leftrightarrow \neg Q, Q \leftrightarrow \neg R, R \leftrightarrow \neg S \vdash P \leftrightarrow S$

3.8 The Strong Completeness of L

A system is *strongly complete with respect to an interpretation* if and only if for every ensemble Γ of wffs and every wff A, if $\Gamma \vDash A$ (on that interpretation) then $\Gamma \vdash A$. The intuitive force of the claim of completeness is then that the rules of the system provide sufficient resources to contruct a proof, for every valid argument, of its conclusion from its premises. Evidently from its definition, strong completeness is the converse of strong soundness. Weak completeness is also the converse of weak soundness:

A system is *weakly complete with respect to an interpretation* if and only if every valid wff is a theorem.

In the previous sections we demonstrated strong soundness and inferred weak soundness from that result. In the following sections we reverse the order for completeness: we demonstrate first weak completeness, and then argue to strong completeness from that result. That is we demonstrate, in the first instance, that every tautology has a proof. Our strategy for doing so is (a) to demonstrate that for any tautology the system L provides us with certain resources; we then (b) give a recipe for constructing a proof of the tautology that exploits specifically those resources. In fact the (a) demonstration depends upon a more general result for wffs in general (not just for tautologies). That result requires some preliminary definitions, the most important of which is that of *the sequent corresponding to a row of a truth-table of a wff A.* The easiest way of defining such a sequent is to say what it has to the left of its $\vdash$ and to the right. Let ρ (rho) be a row of the table of the wff A. Then the sequent corresponding to ρ can be described as follows. On the right, it has A if A receives a 1 in that row; else it has $\neg$A. On the left it has only literals. (A *literal* is a wff that is either an atom or the negation of an atom. Thus, for example, P, $\neg$P are literals; $\neg\neg$P is not.) For each variable V_i having an occurrence in A, the sequent has V_i if V_i receives a 1 in ρ; else it has $\neg V_i$. Thus if A is a wff having occurrences of n distinct atoms $V_1, \ldots, V_n$, and A receives a 1 in ρ, then the sequent corresponding to ρ is

$$W_1, \ldots, W_n \vdash A$$

where W_i is V_i if V_i received a 1 in ρ, is $\neg V_i$ else. If A receives a 0 on row ρ', then the sequent corresponding to ρ' is

$$W'_1, \ldots, W'_n \vdash \neg A$$

where W_i is V_i if V_i received a 1 in ρ, is $\neg V_i$ else. So for example, consider the truth-table for P $\vee$ Q.

	P $\vee$ Q
ρ_1	1 1 1
ρ_2	1 1 0
ρ_3	0 1 1
ρ_4	0 0 0

The sequent corresponding to	ρ_1 is	$P, Q \vdash P \lor Q$;
the sequent corresponding to	ρ_2 is	$P, \neg Q \vdash P \lor Q$;
the sequent corresponding to	ρ_3 is	$\neg P, Q \vdash P \lor Q$;
the sequent corresponding to	ρ_4 is	$\neg P, \neg Q \vdash \neg(P \lor Q)$.

The notion of a sequent corresponding to a row of a truth-table figures in the following lemma, or *intermediate result*:

Lemma 3.1.1

For every wff A, and every row ρ in the truth-table of A, the sequent corresponding to ρ has a proof.

The proof of the lemma takes the form of a strong mathematical induction, this time on the number of occurrences of symbols in A, or as we say more simply, on the length of A. (The wff P has length 1; the wff $\neg$P has length 2; the wff (P $\rightarrow$ (Q $\rightarrow$ P)) has length 9, and so on.) We demonstrate the result for wffs of length 1 (the basis); then we assume that the result holds for all wffs of length < k (H.I.) and demonstrate on that assumption that the result holds for wffs of length k. Since all wffs are of finite length, we can infer that the result holds for all wffs.

The Basis
Assume that A is of length 1. Then A is some atom P_i. The truth-table for A is

$$P$$
$$1$$
$$0$$

The sequent corresponding to row 1 of the table is

$$P \vdash P.$$

The sequent corresponding to row 2 of the table is

$$\neg P \vdash \neg P.$$

Both are provable by the Rule of Assumption.

Inductive Step
Assume that the result holds for every wff A of length $< k$. (H.I.) Let A be a wff of length k. Then A is of one of the following forms: $\neg B$, $B \rightarrow C$, $B \wedge C$, $B \vee C$. We consider these four cases in turn.

Case 1. A is of the form $\neg B$.

Assume the variables of B are $V_1, \ldots, V_n$. The sequents corresponding to the rows of its table are of two kinds:

(a) $W_1, \ldots, W_n \vdash \neg B$

where $\neg B$ received a 1, and

(b) $W_1, \ldots, W_n \vdash \neg\neg B$

where $\neg B$ received a 0.

In the former case (a), B (which is shorter than A ($\neg B$)) received a 0, so by the induction hypothesis,

(a) $W_1, \ldots, W_n \vdash \neg B$

has a proof.

In the latter case (b), B (which is shorter than A ($\neg B$)) received a 1, so by the induction hypothesis,

(b) $W_1, \ldots, W_n \vdash B$

has a proof.

The proof of $\neg A$ is the same proof with the additional line $\neg\neg B$ justified by double negation.

Case 2. A is of the form $B \to C$.

Let $V_1, \ldots, V_n$ be the propositional variables of B
and $V'_1, \ldots, V'_m$ be the propositional variables of C.

There are four subcases corresponding to the four possibilities represented in the rows of the table:

$$
\begin{array}{ccc}
B & \to & C \\
1 & 1 & 1 \\
1 & 0 & 0 \\
0 & 1 & 1 \\
0 & 1 & 0 \\
\end{array}
$$

But B and C are shorter than A $(B \to C)$. Therefore, in the first subcase, there is a proof of

$$W'_1, \ldots, W'_m \vdash C$$

Now write the proof of C and add the line $B \to C$, justified by SI(S) citing sequent **56** $(P \vdash Q \to P)$.

In the second subcase, there is a proof of

$$W_1, \ldots, W_n \vdash B$$

and a proof of

$$W'_1, \ldots, W'_m \vdash \neg C.$$

Now write those two proofs in succession renumbering as required and add the line with $B \wedge \neg C$ as conclusion justified by $\wedge I$ sequent then add the line with $\neg(B \to C)$ as conclusion justified by SI(S) citing Exercise 3.6 (g) ($\dashv$).

In the third and fourth subcases, there is a proof of

$W_1,\ldots,W_n \vdash \neg B.$

Now write that proof and add the line $B \rightarrow C$ as conclusion justified by SI(S), citing **57**

Case 3. A is of the form $B \wedge C$.

Let $V_1,\ldots,V_n$ be the propositional variables of B
and $V'_1,\ldots,V'_m$ be the propositional variables of C.

There are four subcases corresponding to the four rows of the table:

$$
\begin{array}{ccc}
B & \wedge & C \\
1 & 1 & 1 \\
1 & 0 & 0 \\
0 & 0 & 1 \\
0 & 0 & 0 \\
\end{array}
$$

But B and C are shorter than A $(B \wedge C)$. Therefore, in the first subcase, there is a proof of

$W_1,\ldots,W_n \vdash B$

and a proof of

$W'_1,\ldots,W'_m \vdash C.$

Now write those two proofs in succession renumbering as required and add the line with $B \wedge C$ as conclusion justified by $\wedge$I.

In the second subcase, there is a proof of

$W'_1,\ldots,W'_m \vdash \neg C.$

Now write that proof then add the line $\neg B \vee \neg C$, justified by $\vee$I, then add the line $\neg(B \wedge C)$ justified by DeM.

In the third and fourth subcases, there is a proof of

$$W_1,\ldots,W_n \vdash \neg B.$$

In each case, write that proof, then add the line $\neg B \lor \neg C$, justified by $\lor$I, then add the line $\neg(B \land C)$ justified by DeM.

Case 4. A is of the form $B \lor C$.

Let $W_1,\ldots,W_n$ be the propositional variables of B and $W'_1,\ldots,W'_m$ be the propositional variables of C.

There are four subcases corresponding to the four rows of the table:

$$
\begin{array}{ccc}
B & \lor & C \\
1 & 1 & 1 \\
1 & 1 & 0 \\
0 & 1 & 1 \\
0 & 0 & 0
\end{array}
$$

But B and C are shorter than A ($B \lor C$). Therefore, in the first two subcases, there is a proof of

$$W_1,\ldots,W_n \vdash B.$$

In each case, write that proof and add the line $B \lor C$ justified by $\lor$I.

In the third subcase, there is a proof of

$$W'_1,\ldots,W'_m \vdash C.$$

Now write that proof and add the line $B \lor C$ justified by $\lor$I.

In the fourth subcase, there is a proof of

$$W_1,\ldots,W_n \vdash \neg B.$$

and a proof of

$$W'_1,\ldots,W'_m \vdash \neg C.$$

Now write those two proofs in succession renumbering as required and add the line with $\neg B \wedge \neg C$ as conclusion justified by $\wedge I$, then add the line with $\neg(B \vee C)$ as conclusion justified by DeM.

This completes our induction. We have demonstrated the property for wffs of length 1, and have demonstrated that if all wffs of length less than k have the property, then every wff of length k has the property. We conclude that every wff A has the property; that is, for every wff A, and for every row ρ of the truth table of A, the sequent corresponding to ρ has a proof.

3.8.1 The special case of tautologies

Consider the case in which A is a tautology. A receives a 1 in every row of its table. Therefore, in this case, for every row of the truth table, the sequent corresponding to that row has A as its conclusion. If A has n propositional variables, then there are 2^n such sequents available for introduction in the course of a proof of A. We call these sequents *the canonical sequents for A*, and label them $CS1 - CS2^n$. All of these sequents will figure in a *canonical proof for A*.

3.8.2 Weak Completeness: the constructive proof

We will exhibit the proof in the general case first and then consider an example.

Let the propositional variables of A be $V_1,\ldots,V_n$. Then the first 3n lines of the proof of A will be

(1)	$V_1 \vee \neg V_1$	LEM
.		
.		
(n)	$V_n \vee \neg V_n$	LEM

| n+1 | (n+1) | V_1 | A (∨E) |
| n+2 | (n+2) | $\neg V_1$ | A (∨E) |

.

| 3n-1 | (3n-1) | V_n | A (∨E) |
| 3n | (3n) | $\neg V_n$ | A (∨E) |

The next section will consist of 2^n lines each having A as its conclusion, and justified by reference to CS1 – CS2^n.

| n+1,n+3,..,3n-1 | (3n+1) A | n+1,n+3,..,3n-1,CS1 |

.

| n+2,n+4,..,3n | (3n+2^n) A | n+2,n+4,..,3n,CS2^n |

The first of the n final sections will consist of 2^{n-1} lines each having A as its conclusion, and justified by ∨E citing line (1), and the lines where its disjuncts are assumed. The second of these final sections will consist of 2^{n-2} lines each justified by ∨E and citing line (2) and the lines where its disjuncts are assumed… The nth final section will consist of 2^0 (*i.e.* 1) line, the last line of the proof, justified by ∨E and citing line (n), the lines where its disjuncts are assumed, and the penultimate and antepenultimate lines of the proof.

We are now in a position to assert that the system L is weakly complete with respect to the given interpretation.

3.8.3 Some examples of canonical proofs

Consider the tautology in two variables

$$P \to P \vee Q.$$

Its canonical proof is as follows:

| (1) | $P \vee \neg P$ | LEM |
| (2) | $Q \vee \neg Q$ | LEM |

3	(3)	P	A (∨E)
4	(4)	¬P	A (∨E)
5	(5)	Q	A (∨E)
6	(6)	¬Q	A (∨E)
3,5	(7)	P → P ∨ Q	3,5,CS1
3,6	(8)	P → P ∨ Q	3,6,CS2
4,5	(9)	P → P ∨ Q	4,5,CS3
4,6	(10)	P → P ∨ Q	4,6,CS4
5	(11)	P → P ∨ Q	1,3,7,4,9,∨E
6	(12)	P → P ∨ Q	1,3,8,4,10,∨E
	(13)	P → P ∨ Q	2,5,11,6,12,∨E

As a second example, consider the tautology in three propositional variables:

$$A = (P \to (Q \lor R)) \to ((P \to Q) \lor (P \to R))$$

which, for the sake of brevity, we shall call A. This represents a slightly more complicated case than the previous example, but should serve to illustrate the general form that such a canonical proof takes.

Proof of A

	(1)	P ∨ ¬P	LEM
	(2)	Q ∨ ¬Q	LEM
	(3)	R ∨ ¬R	LEM
4	(4)	P	A (∨E)
5	(5)	¬P	A (∨E)
6	(6)	Q	A (∨E)
7	(7)	¬Q	A (∨E)
8	(8)	R	A (∨E)
9	(9)	¬R	A (∨E)
4,6,8	(10)	A	4,6,8,CS1
4,6,9	(11)	A	4,6,9,CS2
4,7,8	(12)	A	4,7,8,CS3
4,7,9	(13)	A	4,7,9,CS4

5,6,8	(14) A	5,6,8,CS5
5,6,9	(15) A	5,6,9,CS6
5,7,8	(16) A	5,7,8,CS7
5,7,9	(17) A	5,7,9,CS8
6,8	(18) A	1,4,10,5,14,∨E
6,9	(19) A	1,4,11,5,15,∨E
7,8	(20) A	1,4,12,5,16,∨E
7,9	(21) A	1,4,13,5,17,∨E
8	(22) A	2,6,18,7,20,∨E
9	(23) A	2,6,19,7,21,∨E
	(24) A	3,8,22,9,23,∨E

The above proof is not only the canonical proof for

$$(P \rightarrow (Q \vee R)) \rightarrow ((P \rightarrow Q) \vee (P \rightarrow R))$$

It is also the canonical proof for any tautology A in the three propositional variables P, Q and R.

3.8.4 Strong Completeness

We now show how to extend the proof of weak completeness of L to a proof of strong completeness. The method involves the notion of *the conditional corresponding to a sequent*, which is defined by the following:

The conditional corresponding to a sequent is the conditional formed by iterating the following operation:

> *Conditionalize the conclusion of the sequent on the rightmost premiss, then delete that premiss*

until no premisses remain.

For example, let the sequent be A, B, C ⊢ D. In the first application, we conditionalize D on C and then delete the premiss C to give

A, B ⊢ (C → D). In the second application, we conditionalize (C → D) on B and then delete the premiss B to give A ⊢ (B → (C → D)). In the final application we conditionalize (B → (C → D)) on A, and delete the premiss A to give

$$A \rightarrow (B \rightarrow (C \rightarrow D))$$

Now by completeness we have shown that every tautology is an L-theorem. It follows from that result that every tautology which is the corresponding conditional of a sequent is provable. We can show that every entailment has a proof, by indicating how the proof is to be constructed from that of its corresponding conditional. Suppose that

$$A \rightarrow (B \rightarrow (C \rightarrow D))$$

is a tautology. By our weak completeness result, we know that that wff has a proof. Let its proof be

.
.
.

(n)   A → (B → (C → D))   <rule> etc..

Then the proof of its corresponding sequent

$$A, B, C \vdash D$$

will be obtained by adding A, B and C as initial assumptions to the proof of A → (B → (C → D)), renumbering, and then applying MPP as follows:

1	(1)	A	A
2	(2)	B	A
3	(3)	C	A
	.		
	.		

	(n+3)	$A \to (B \to (C \to D))$	\<rule\> etc.
1	(n+4)	$B \to (C \to D)$	MPP 1, n+3
1,2	(n+5)	$C \to D$	MPP 2, n+4
1,2,3	(n+6)	D	MPP 3, n+5

With this we conclude the main demonstration of this section. We have now shown that the system L is strongly complete with respect to the interpretation given it.

This extension of the demonstration of weak completeness to a demonstration of strong completeness prompts an observation and a general remark. The extension is possible because the system L satisfies a condition that it shares with some other, but not all systems of logic. This is the equivalence given by the biconditional

$$\text{For every wff A and every wff B, } A \vdash B \text{ iff } \vdash A \to B$$

and usually referred to the *The Deduction Theorem*. Not every formal system is one for which the Deduction Theorem holds, and in some systems in which it does not, strong completeness cannot be demonstrated. In other such systems, strong completeness must evidently be demonstrated by other means.

3.8.5 Consequences of completeness

We are now in a position to demonstrate for a given sequent that it has a proof, even if we find ourselves, inexplicably, unable to prove it, or even to produce the canonical proof for its corresponding conditional. Citing the Completeness Metatheorem, we can establish that a sequent has a proof by demonstrating that its premises entail its conclusion. Practically, we might try hypothetically assigning 1 to each of its premises and 0 to its conclusion, then computing the values that that hypothetical assignment dictates. If the hypothesis leads to an assignment of conflicting values to distinct occurrences of the same atom, then we can conclude that no such assignment is possible, that any assignment that makes all of the premises true must also make the conclusion true, that is that the premises entail the conclusion. It follows, by the Completeness Metatheorem, that the sequent has a proof in L.

The reader will have noticed some correspondences between the uses of soundness and those of completeness. Citing soundness, we can demonstrate non-theoremhood of any non-valid wff, or the unprovability of any non-valid sequent. Citing completeness, we can demonstrate the provability of any valid wff, or the provability of any valid sequent. One final correspondence will finish this section. We have seen that it follows from the soundness of L that L is also consistent. It follows from the completeness of L that L is also *maximal*. A system is *maximal* iff the addition of any new wff as a theorem yields an inconsistent system, that is, one in which an inconsistent wff is a theorem.

The set of theorems of L lies within an important class of sets of wffs: those that are said to be maximal L-consistent. To sum up that notion, a set Σ of wffs is *maximal L-consistent* iff

(a)   Σ is L-consistent (no contradiction is L-provable from Σ) and

(b)   For any wff A that is not a member of Σ, the set resulting from adding A to Σ is L-inconsistent.

Maximal L-consistent sets are not a rarity. It is straightforward to demonstrate that every L-consistent set can be extended to a maximal L-consistent set. It is also straightforward to demonstrate that the set of such sets is uncountably large. However, the set of L-theorems has the additional property that it is closed under uniform substitution. It is the only maximal L-consistent set that is also so-closed.

A system is derivatively called maximal, if its set of theorems is maximal. It is alternatively said to be a *Post complete*[4] system. Not every complete system is Post complete, but L is easily shown to be such a system.

4. After Emil L. Post (1897–1954) who discussed the property.

Corollary 3.0.3

L is Post complete.

Proof:
Let **A** be a wff that is not a theorem of L. Consider the system LA which has all the rules of L, but which also lets us introduce **A** into a proof by TI and any substitution instance of **A** into a proof by TI(S). Since **A** is not a theorem of L, it follows by the completeness metatheorem that **A** is not a valid wff, whence it follows that **A** has an inconsistent substitution instance. (Let ρ_i be the row of the truth-table for **A** in which **A** receives a 0. For each propositional letter, P_i that receives a 1 in that row uniformly substitute $P_i \vee \neg P_i$. For each propositional letter, P_i that receives a 0 in that row uniformly substitute $P_i \wedge \neg P_i$. The resulting substitution instance must have the value 0 on every assignment of truth-values to its propositional letters, and is therefore inconsistent.) But that wff has a one-line proof in LA, by TI(S), **A**.   Q.E.D.

Exercise 3.14

Calculate the length of the canonical proof of a tautology in n distinct propositional variables.

Quantificational Logic 1

4.1 The Inadequacy of Propositional Forms

Recall the remarks with which this presentation began. Logicians have traditionally been interested in arguments the goodness or badness of which turns upon the presence in them of certain vocabulary. In the previous chapters we have seen something of the logical theory that has grown out of logicians' interest in such words as *or, and, not,* and *if.* We have already noted that that vocabulary is inadequate to account for the goodness of some types of good argument. We have said that much of our dictionary-definable vocabulary (that, say, of colour, or spatial relationships, for example) impose restrictions on inference that could be presented as sets of rules. With perseverence, we might construct formal systems along the lines of L for such vocabulary. The difficulty for such systems is that they would likely not find wide enough fields of application to justify the bother. They have at least not so far captured many formalists' interest. However, there remain a family of good arguments whose crucial vocabulary is both general enough in its applications to interest formalists, and inexpressible in the language of L. Consider the argument:

All logic students are impressive debaters;
Olive is a logic student;
therefore, Olive is an impressive debater.

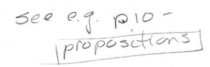

If we are restricted to the language of L, then the best representation of the argument will be the sequent:

$$P, Q \vdash R$$

which, by the Soundness Metatheorem, we know to have no proof in L. Nevertheless, intuitively the argument has a valid form, a form which it shares with the argument:

> All politicians are scrupulously honest;
> Jeffrey Archer is a politician;
> therefore, Jeffrey Archer is scrupulously honest.

What these arguments have in common is that their goodness turns upon the behaviour of what some linguists call *quantified noun phrases* ('all logic students', 'all politicians') and upon attributions to individuals of the properties introduced in those quantified noun phrases. We need new logical vocabulary and new rules of inference governing *that* vocabulary if those arguments are to be licensed by a formal system that includes L. To see what new vocabulary is required, we investigate the common form of the two previous arguments. The first premiss of each may be taken to assert of every object in the universe, that if it has one property (in the former case, being a logic student; in the second, being a politician), then it has another property (being an impressive debater; being scrupulously honest.) The second premiss of each asserts of some individual that it has the former property. The third asserts of the same individual that it has the other property. The intuitive correctness of the argument is, however, independent of the properties and the individual selected. Thus we could represent their common correct form as:

> It is true as regards any individual, that if it has property F, then it has property G;
> The individual m has the property F;
> therefore, the individual m has the property G.

Here, m is a *proper name* and F and G are *predicate symbols*.

In our language we will write: 'Fm' to represent "the individual m has the property F" and 'Gm' to represent "the individual m has the property G." Here 'm' (or 'l' or 'n') is playing the role of a proper name. We also need some vocabulary that will behave somewhat like the pronominal vocabulary of natural language to represent the two occurrences of "it" in the first premiss. For this purpose, we will use x, y, z. Thus 'Fx' could be understood informally as representing a sentence such as "It has the property F." 'x', 'y', 'z', and so on, are called *individual variables*, the intuitive idea being that, like pronouns, they vary over individuals, making reference to no particular individual independently of contextual, gestural, or other specific fixing. We continue to use the $\to$ of the language of L to represent the "if...then...", but we now need some way of saying that the conditional:

$$(Fx \to Gx)$$

is true of every individual x. For this, we use '(x)', *i.e.* 'x' enclosed in parentheses. '(x)', (or '(y)', '(z)' *etc.* is called a *universal quantifier*, and is read: 'It is true as regards every individual x that . . .' or more simply 'For every x . . .' or 'For all x . . .'. Thus the form of the first premiss may be represented by:

$$(x)(Fx \to Gx)$$

and the form of the argument as a whole by:

$$(x)(Fx \to Gx)$$
$$Fm$$
$$\text{therefore, Gm}$$

or as the sequent:

$$(x)(Fx \to Gx), Fm \vdash Gm.$$

So far, so good. However, consider the argument:

All logic students are impressive debaters;
Some business students are logic students;
therefore, some business students are impressive debaters.

which is also intuitively correct.

We require some way of expressing the second premiss and the conclusion, in particular, some way of representing the role that the quantified noun phrase 'some business students' is playing in the argument.

Now each of these may be regarded as asserting that there are individuals which combine two properties: in the first case, the property of being a business student and the property of being a logic student, and in the second, the property of being a business student and that of being an impressive debater. Minimally, each can be regarded as saying that *there is at least one individual such that* it has the first property and it has the second. As before, the pronoun "it" can be represented by an individual variable 'x'. To represent the italicized expression we introduce the prefix '$(\exists x)$' which we read 'There is at least one individual x such that . . .', or more simply 'There exists an x such that . . .' or 'For some x, . . .'. '$(\exists x)$' is called an *existential quantifier*. The argument is then represented by the sequent:

$$(x)(Fx \to Gx), (\exists x)(Hx \wedge Fx) \vdash (\exists x)(Hx \wedge Gx).$$

The repetition of the individual variable of the quantifier in the remainder of the formula has the effect of fixing the grammatical reference and making it precise. So, as in English usage by which in the sentence

   That individual has the property that it is white and that it is flat

we understand that whatever individual is being spoken of, both occurrences of "it" refer to the same one, so in

$$(x)(Fx \to Gx)$$

which we read: 'For every (individual) x, if Fx, then Gx', the repetition of x makes it explicit that the bearer of property F is the same as the bearer of property G. Similarly, in the sentence

$$(\exists x)(Hx \wedge Fx)$$

which we read: 'There exists an (individual) x such that Hx and Fx' the repetition of x makes it explicit that the bearer of property H is the same as the bearer of property F.

4.1.1 n-place predicates

With just the vocabulary that we have so far, we can represent the forms of an impressive variety of English sentences. Just to consider two simple forms not yet mentioned. The sentence

No women are immortal

is understood as being equivalently expressed by 'All women lack the property of being immortal' and its form is therefore is represented by

$$(x)(Fx \rightarrow \neg Gx).$$

The form of the sentence

Not all misers are Phrygians

could of course be represented as

$$\neg(x)(Fx \rightarrow Gx)$$

Some

but the sentence can also be understood as equivalently expressed by 'Some misers are not Phrygians' and its form accordingly represented by

$$(\exists x)(Fx \wedge \neg Gx). \quad \longleftarrow \text{ preferred}$$

For the present, quantificational sentences that begin with quantifiers hold certain practical advantages over quantificational sentences that begin with negations, so the second representation is preferred to the first.

Now in both of these examples, and all of the examples we have considered so far, the properties mentioned, whiteness, flatness, immortality, and so on, have been *1-place properties*, that is, properties which are exhibited by single individuals. But some properties are relational in character and can be manifested by two or more individuals. The property of motherhood and betweenness can be regarded as

[handwritten: 2 or more place properties = n place predicates]

properties exhibited by pairs and by triples of individuals respectively. Such properties we represent by a predicate symbol followed by the required number of proper names or individual variables, as in

Tiffany is the sister of Kimberley

Fmn

Miles is the lover of Laurel's fiancé

Fnl ∧ Lmn   *[handwritten: Fiancé level; n = Laurel's fiancé]*

Miles is the lover of at least one of Laurel's fiancés

(∃x)(Fxl ∧ Lmx)

Graham is between Laurel and Candice

Fmnn′ *[handwritten: ?]*

Everyone has a mother

(x)(Fx → (∃y)Gyx).

Here the force of 'everyone' is taken to mean 'everything that is a person', and 'F' is introduced to represent the property of being a person.

Every boy loves some girl

(x)(Bx → (∃y)(Gy ∧ Lxy))

As to this sentence, there might be some debate as to whether it claims that there is some girl that every boy loves, that is, whether it ought rather to be represented by

(∃x)(Gx ∧ (y)(By → Lyx)).

There is no general answer to the question as to which it means. In natural language, the more natural question is whether we can *force* a reading in one way or another by the devices of speech, the so-called *prosodic* presentation of the sentence, the combination of stress, pitch contour, and lengthenings of sounds. Imagine someone voicing the suspicion that young what's-his-name's general lethargy and listlessness might be a symptom of infatuation with some or other young woman, and replying that the suspicion was not unreasonable, since, after all, every boy loves some girl. Now pay particular attention to the way that you would find yourself uttering the sentence. Those prosodic characteristics, so important to linguistic exchanges, cannot be represented in any notation that we make use of here. In fact the methodology that such a study would require might make only incidental use of anything that logic has to offer. These sentences provide useful examples for becoming acquainted with the language of quantifiers, but probably in the end do not represent the most important application of a quantificational formal system. We can of course make distinctions syntactically, and it might be a worthwhile exercise to ponder or to explore the limitations of the range of syntactic distinctions that we can make in quantificational language. But those questions lie far beyond the scope of this text. Here we note that the two previous representations are distinct from one another, and that both are distinct from the representation of

Some boy loves every girl

$(\exists x)(Bx \land (y)(Gy \rightarrow Lxy))$.

Now we have said that 2-place predicates represent properties that are exhibited by pairs of individuals. As far as it goes, that claim is true. In general n-place predicates are understood as representing properties exhibited by n-tuples of individuals. But this should not be understood as involving the requirement that all the individuals of the n-tuple must be distinct. To be sure, if M is understood as representing the 2-place predicate "is the mother of", then Mnn will have to take the value 0 on that interpretation, but the constructions such as Mnn in which a proper name or individual variable appears in both places is required if we are to express the sentence

Norman is not his own mother

¬Mnn

or the sentence

No one is his own mother

(x)(Px → ¬Mxx)

and we want to be able to express such sentences as

Some boy loves himself

(∃x)(Bx ∧ Lxx)

and we want to be able to distinguish the two readings of the sentence

No one respects anyone who does not respect himself

(x)(y)(Px ∧ Py → (¬Rxx → ¬Ryx))

(x)(y)(Px ∧ Py → (¬Rxx → ¬Rxy)).

Again notice how your prosodic presentation of the sentence would vary accordingly as you intend the one reading and intend the other.

Finally we note that natural language usually provides a wide variety of more or less equivalent constructions. In English, for example, we need not say 'Everyone who has a mother has a parent', we could say instead

Only those who have a parent have a mother

(x)(Fx → ((∃y)Myx → (∃y)Pyx)).

In general, the better we speak a language, the more such forms we have at our disposal, and the subtler the distinctions we are capable of attending to.

Exercise 4.1

For each of the following English sentences, use **Simon**'s formulation editor to give a representation of it in the quantificational notation introduced so far. In each case the number of proper names and number of predicate symbols suggested is a guide to the number of such symbols required. In some cases, all predicate symbols will have one-place occurrences; in others some will have two-placed occurrences. In almost all cases, the predicate symbols given are available on the virtual keyboard of the formulation editor.

(a) Porky is a Pig   (Fx: x is a pig; Fm
                       m: Porky)

(b) Michael is not a socialist   (Fx: x is a socialist; $\neg Fm$
                                   m: Michael)

(c) Some socialists are under-fed   (Fx: x is a socialist; $(\exists x)\, (Fx \wedge Gx)$
                                      Gx: x is underfed)

(d) Only artists understand life   (Fx: x is an artist;
                                     Gxy: x understands y;
                                     l: life)

(e) No dog likes vegetables   (Fx: x is a dog;
   $(x)(y)\,((Fx \wedge Hy) \supset \neg Gxy)$ Gxy: x likes y;
                                Hx: x is a vegetable)

(f) No dog likes mouldy meat   (Fx: x is a dog;
                                 Gxy: x likes y;
                                 Hx: x is mouldy;
                                 Jx: x is meat)

(g) Brutus killed Caesar   (Fxy: x killed y;
        Fmn   m: Brutus;
                                             n: Caesar)

(h) Brutus killed someone   (Fxy: x killed y;
                                             n: Brutus;
                                             Gx: x is a person)

(i) Someone killed someone   (Fxy: x killed y;
                                             Gx: x is a person)

(j) Someone killed himself   (Fxy: x killed y;
                                             Gx: x is a person)

(k) No one killed himself   (Fxy: x killed y;
                                             Gx: x is a person)

(l) Someone helped everyone   (Fxy: x helped y;
                                             Gx: x is a person)

(m) Everyone helped some-   (Fxy: x helped y;
     one   Gx: x is a person)

(n) Someone was helped by   (Fxy:x helped y;
     everyone   Gx: x is a person)

(o) Some socialists like no   (Fx: x is a socialist;
     cuts   Gxy: x likes y;
                                             Hx: x is a cut)

(p) Some dogs like all cuts   (Fx: x is a dog;
                                             Gxy: x likes y;
                                             Hx: x is a cut)

(q) All dogs like some cuts   (Fx: x is a dog;
                                             Gxy: x likes y;
                                             Hx: x is a cut)

(r) Every dog has his day

(Fx: x is a dog;
Gx: x is a day;
Hxy: x has y;
Jxy: y is x's)

(s) Only the brave deserve the fair

(Fx: x is brave;
Gxy: x deserves y;
Hx: x is fair)

(t) The brave deserve only the fair

(Fx: x is brave;
Gxy: x deserves y;
Hx: x is fair)

(u) All that glisters is not gold

(Gx: x glisters;
Hx: x is gold)

(v) Everyone loves a lover

(Fx: x is a person;
Gxy: x loves y)

(w) No one loves a stinker

(Fx: x is a person;
Sx: x is a stinker;
Hxy: x loves y)

(x) If one makes a bed, one lies in it

(Fx: x is a person;
Gx: x is a bed;
Hxy: x makes y;
Jxy: x lies in y)

(y) The good is better than the bad

(Fx: x is good;
Gx: x is bad;
Hxy: x is better than y)

(z) The best is better than the worst

(Fxy: x is better than y)

(aa) The best is better than the worst

(Fxy: x is worse than y)

Hints

In (r), we must ponder why we are given four predicate symbols. Two of them will have one-place occurrences; two will have two-place occurrences.

In (v) we will use 'Hxy' to represent 'x loves y'. But now we will have to sort out how best to represent the property of being a lover. To be a lover is to be a lover *of* something.

In (z), we must determine how best to deploy the sole predicate symbol 'F' to produce an approximate quantificational representation. Evidently its occurrences will have to two-place occurrences to accommodate 'is better than'. The tricky part will be to capture as nearly as possible 'the best' and 'the worst'. Bear in mind that the worst of a kind is not that than which everything of the kind is better; that would require the worst to be better than itself. Parallel remarks apply to the best of a kind. Bear in mind also that our language is as yet insufficiently expressive to represent the worst as that which is worse than everything *else*. So we are forced to represent the worst as that which is not better than anything. And the best?

4.2 Quantificational rules

4.2.1 Introduction

The four quantificational rules comprise an introduction and an elimination rule for each quantifier. They can be conveniently understood as generalizations of the introduction and elimination rules for $\wedge$ and $\vee$, the rules for the universal quantifier being akin to those for $\wedge$, and those for the existential quantifier being akin to those for $\vee$. The reason for the kinship is that the quantifiers themselves can be regarded as generalizations of conjunction and disjunction. This point deserves a little expansion.

Imagine that we are making observations about some finite group of objects, a set of natural numbers, say, or a gathering of friends. Since there are only finitely many of them, if we want to say that they share some property, say that they are all less than fifty years old (in either example), we can either use a quantified noun phrase: 'All of them are under fifty' or we can conjoin claims made about each of them individually: 'Arabella is under fifty and Fred is under fifty and . . . and Zebediah is under fifty.' From either of the two forms of the claim our audience would be entitled to infer of anyone in the group that that person or that number of persons are under fifty. In the latter case the inference would be licensed by the ordinary understanding of *and*; in the former case, by the ordinary understanding of *all*. The same understanding would entitle us to infer either the conjunction or the *all*-sentence from an exhaustive set of separate but uniform claims for the individuals in the room. Natural language also allows us to place an *and*-list of items at the head of the attribution, as

> Arabella and Fred and . . . and Zebediah are under fifty

a construction even more closely paralleling the quantified noun phrase construction.

Corresponding remarks apply to claims couched in the language of *or* and the language of *some*. There are only practical differences between the claim that at least one of the objects is odd, and the exhaustive *or*-construction out of individual attributions of oddness to members of the group. And as in the case of *and* natural language permits us to put an *or*-list at the head of the attribution

> Arabella or Fred or . . . or Zebediah is under fifty

so we can think of the existentially quantified noun phrase as standing in for an *or*-list of items. Either sentence is justifiably inferred from such an attribution to any single member of the group. And from either sentence we are justified in making only those inferences that are indifferent as to which member of the group has the property in question. Anything that we take to follow must be such as to follow whichever member of the group has the property.

So far, at the level of ordinary language at least, one might suppose that there is really no reason other than the convenience of greater brevity to introduce quantificational language into speech at all. However, two kinds of cases present obstacles. The first are the cases in which the objects are practicably indistinguishable, as 'all the bees in that swarm' or for some other reason do not have and cannot be given names. Second, some such device is required for cases in which the class of objects to all or some of which the uniform attribution is being made is infinitely large. In this case there is no question of replacing the quantifier with an exhaustive conjunctive or disjunctive sentence, and no list that trails off into suspension dots will make the range of the uniform attribution explicit. So, particularly for applications in the formal sciences, we need quantifiers in our logical theory. But the namelessness cases (some bee in that swarm) and the infinite case do present a difficulty for the formulation of an introduction rule for the universal quantifier and the elimination rule for the existential quantifier. The parallel with ∧ would seem to require, in this case, infinitely many input sentences; the parallel with ∨ would seem to require infinitely many subproofs.

The difficulty is surmountable because of the uniformity of the attributions that brings the list and the quantified noun phrase into play in natural language. For the introduction rule for the universal quantifier, we can prove the attribution for an *arbitrary* member of the group, that is, we can prove it for an item about which we make no assumptions other than that it is a member of the group. For the elimination rule for the existential quantifier we can prove that the conclusion follows from the attribution to an arbitrary member of the group, again, a member about which we make no assumptions other than that it is a member of the group. Those intuitions have guided the development of the rules, which would be practically useless if they were not shaped by their proposed application. However, we must remember that our rules are rules for the constructions of proofs, and must mention only what can be discovered by inspection of sentences already inscribed. We shall therefore not officially mention arbitrary individuals, but rather *arbitrary names*, that is names that (though we may think of them as being essentially not names of any particular individual) are simply special notation available for certain proof-theoretic purposes, and satisfying such further requirements in the proof as the particular rule stipulates.

4.2.2 The Rule of Universal Elimination (UE)

We will be in a position to give a more precise account of all of the rules a little later. At the moment, it is sufficient to use the rule analogously with our use of ∧-elimination. That is, if a sentence having a universal quantifier (v) as its connective of longest scope occurs as a line of a proof, the rule UE will permit us to write down as a line, the result of removing the quantifier, and replacing all remaining occurrences of its individual variable, v with any name whatsoever. At the moment we have only proper names, but the rule will permit substitutions of arbitrary names as well. The required assumptions are those of the input line. In its simplest application, the rule would provide a proof of the sequent

62   (x)Fx ⊢ Fm

1	(1)	(x)Fx	A
1	(2)	Fm	1,UE

UE permits the introduction of an arbitrary name.

← just the name of an individual in the group.

and, more interestingly of the sequent on page 109.

63   (x)(Fx → Gx), Fm ⊢ Gm

1	(1)	(x)(Fx → Gx)	A
2	(2)	Fm	A
1	(3)	Fm → Gm	1,UE
1,2	(4)	Gm	2,3,MPP

4.2.3 The Rule of Universal Introduction (UI)

The intuitive basis of this rule is as it was outlined in the introductory remarks. Its use, in essentially the form to be described, is long established in mathematical practice. Suppose that we want to prove that every isosceles triangle has equal base angles. The standard form of the proof begins 'Let ABC be a (or 'an arbitrary' or 'any') triangle with AB = AC.' The proof is given that angle ABC = angle ACB, and then, because ABC was an *arbitrary* isosceles triangle, that is, a triangle about which nothing was assumed except that it was isosceles, it is inferred that

arbitrary name

every isosceles triangle has equal base angles. Our application of the practice uses the device of the *arbitrary name*. For this purpose we reserve a, b, c, d, a′, b′ and so on. With certain reservations UI permits us to write down a universally quantified sentence on the basis of the corresponding attribution involving an arbitrary name, provided no assumptions are required in which the arbitrary name appears. This extra requirement captures at a purely proof-theoretic level the historical notion of arbitrariness, the notion that no special assumptions have been made about the individual in question. As we remarked earlier, UE also permits the introduction of an arbitrary name, as in the proof of

64 $(x)Fx \vdash Fa$

1	(1)	$(x)Fx$	A
1	(2)	Fa	1,UE

In the course of constructing proofs and in devising strategies for them, practical considerations will guide our choice between arbitrary names and proper names in applications of UE. In particular, in some cases where we intend to introduce a universal quantifier later in a proof, we will elect to use an arbitrary name rather than a proper name. Consider the proof of

65 $(x)(Fx \rightarrow Gx), (x)(Gx \rightarrow Hx) \vdash (x)(Fx \rightarrow Hx)$

1	(1)	$(x)(Fx \rightarrow Gx)$	A
2	(2)	$(x)(Gx \rightarrow Hx)$	A
3	(3)	Fa	A(CP)
1	(4)	$Fa \rightarrow Ga$	1,UE
1,3	(5)	Ga	3,4,MPP
2	(6)	$Ga \rightarrow Ha$	2,UE
1,2,3	(7)	Ha	5,6,MPP
1,2	(8)	$Fa \rightarrow Ha$	3,7,CP
1,2	(9)	$(x)(Fx \rightarrow Hx)$	8,UI

General
↓
arbitrary name
↓
General.

Notice that were we presenting a demonstration here, expressing the lines of the proof in mathematical English, we would have been justified after line (8) in saying 'But 'a' was arbitrary.' The reason is that neither of the assumptions required at line 8 has any occurrences of the arbitrary name 'a'. In working through the next proof, observe that the requirement is also satisfied.

66 (x)(Fx → Gx), (x)Fx ⊢ (x)Gx

1	(1)	(x)(Fx → Gx)	A
2	(2)	(x)Fx	A
1	(3)	Fa → Ga	1,UE
2	(4)	Fa	2,UE
1,2	(5)	Ga	3,4,MPP
1,2	(6)	(x)Gx	5,UI

Exercise 4.2

In this exercise, we will represent arguments as sequents and give proofs for those sequents. All of this can be done within **Simon**. First, for each of the following arguments, use **Simon**'s formulation editor to represent each of its sentences as a quantificational wff. Make sure that you formulate the conclusion last. The resulting formulations will be added automatically to the formulation catalogue (the window to the left of the formulation editor). Then highlight all of the formulations of the premises and conclusion, right click, and then choose 'Encapsulate selections as a sequent'. **Simon** will transfer your formulations to the proof editor's catalogue and give it a name. (You can change the name if you wish.) Then use **Simon**'s proof editor to construct a proof of the sequent using the rules of L together with UE and UI as required.

(a) Socrates is a philosopher. All philosophers are a little odd. Therefore Socrates is a little odd.

(b) Socrates is secretly vain. No philosopher is secretive about his vanity. Therefore Socrates is not a philosopher.

(c) Spike is no logician. Only logicians properly mind their P's and Q's. Therefore Spike does not properly mind his P's and Q's.

(d) All male cardinals are red. Pauli is not red. Pauli is male. Therefore Pauli is not a cardinal.

(e)   All Frenchmen except Parisians are kindly. Jacques is a Frenchman; Jacques is not kindly. Therefore Jacques is a Parisian.

Exercise 4.3

Using UE and UI together with the rules of L, construct proofs for the following sequents:

(a) $(x)(Fx \rightarrow Gx), (x)(Gx \rightarrow \neg Hx) \vdash (x)(Fx \rightarrow \neg Hx)$
(b) $(x)(Fx \rightarrow \neg Gx), (x)(Hx \rightarrow Gx) \vdash (x)(Fx \rightarrow \neg Hx)$
(c) $(x)(Fx \rightarrow Gx), (x)(Hx \rightarrow \neg Gx) \vdash (x)(Fx \rightarrow \neg Hx)$
(d) $(x)(Gx \rightarrow \neg Hx), (x)(Hx \rightarrow Gx) \vdash (x)(Fx \rightarrow \neg Hx)$
(e) $(x)(Fx \rightarrow Gx) \vdash (x)Fx \rightarrow (x)Gx$
(f) $(x)(Fx \vee Gx \rightarrow Hx), (x)\neg Hx \vdash (x)\neg Fx$

Exercise 4.4

For each of the following arguments, say which of the above sequents can be taken to represent it.

(a)   No Germans are Frenchmen. All Bavarians are German. Therefore, no Frenchmen are Bavarians.
(b)   No Frenchmen are beer drinkers; all Bavarians are beer drinkers. Therefore, no Frenchmen are Bavarians.
(c)   All Bavarians are beer drinkers. No Frenchmen are beer drinkers. Therefore no Frenchmen are Bavarians.
(d)   All Germans are patriots. No patriots are insincere. Therefore no Germans are insincere.

4.2.4 The Rule of Existential Introduction (EI)

As universal elimination is akin to ∧-elimination, so existential introduction is akin to ∨-introduction, and most easily understood by that connection. Its intuitive origins lie in the fact that if a property holds of a particular object, then there is an object of which the property holds. Thus the rule permits a proof of either of the simple sequents

67 Fm ⊢ (∃x)Fx

1	(1)	Fm	A
1	(2)	(∃x)Fx	1,EI

68 Fa ⊢ (∃x)Fx

1	(1)	Fa	A
1	(2)	(∃x)Fx	1,EI.

It also permits a proof of

69 (x)Fx ⊢ (∃y)Fy

1	(1)	(x)Fx	A
1	(2)	Fa	1,UE
1	(3)	(∃y)Fy	2,EI.

4.2.5 The Rule of Existential Elimination (EE)

Again, existential elimination is best understood along the lines of ∨-elimination. Where ∨E permits us to write down as proved from the disjunction A ∨ B, any sentence that has been proved from A and proved from B, EE permits us to write down as proved from an existentially quantified sentence, any sentence that has been proved from an arbitrary instance. So for example, it would let us write as proved from (∃x)Fx, any sentence C that has been proved from Fa. The intuitive basis for the rule is this: (∃x)Fx requires that some object has the property F. In proving C from Fa, we are proving that whichever object has the property, C is provable. We can then assert C, given only that (∃x)Fx. The required assumptions of C are the assumptions required for proving C from Fa, except Fa itself. The justification will cite three lines: the existentially quantified sentence, and the beginning and end of the proof of C from Fa. There are some restrictions to observe in this proof. The proof of C must not require any assumptions about a other than Fa, and the arbitrary name a must not appear in C. It is as though in a less formally expressed mathematical demonstration, we have the line: *there is an object having the property F.* We then continue: *Let that object be a.* We then go on to prove some conclusion. We do not want that conclusion to depend upon a's being any particular object about which something extraneous is already known, and we don't want to conclude that 'a' is really what the object is called. 'a' was just a label that we attached to the object for the sake of the proof. As an example, consider the proof of

70 $(x)(Fx \to Gx), (\exists x)Fx \vdash (\exists x)Gx$

1	(1)	$(x)(Fx \to Gx)$	A
2	(2)	$(\exists x)Fx$	A
3	(3)	Fa	A (EE)
1	(4)	$Fa \to Ga$	1,UE
1,3	(5)	Ga	3,4,MPP
1,3	(6)	$(\exists x)Gx$	5,EI
$C = $ 1,2	(7)	$(\exists x)Gx$	2,3,6,EE.

Notice that both the conditions on C in the statement of the rule are satisfied by line 6. First, no assumptions are required by line 6 except the assumption at line 3, and second, the arbitrary name a does not appear in line 6. By contrast, the following would *not* be a proof of that sequent:

1	(1)	$(x)(Fx \to Gx)$	A
2	(2)	$(\exists x)Fx$	A
3	(3)	Fa	A (EE)
1	(4)	$Fa \to Ga$	1,UE
1,3	(5)	Ga	3,4,MPP
1,3	(6)	Ga	2,3,5,EE✗
1,2	(7)	$(\exists x)Gx$	6,EI

because the arbitrary name a appears in line 5, which purports to be the conclusion C.

As a further illustration of the restrictions and the non-proofs which their observance precludes, consider the following argument, its correct representation as a sequent and some attempts at a proof of the sequent. The argument is this:

Every philosopher is benevolent. There are some folk, who if they are benevolent, are smug. Therefore every philosopher is smug.

It is correctly represented by the following sequent:

$(x)(Px \rightarrow Bx), (\exists x)(Fx \wedge (Bx \rightarrow Sx)) \vdash (x)(Px \rightarrow Sx).$

What follows cannot be called a proof, but, suggesting Pinocchio, we could perhaps call it a prooof.

1	(1)	$(x)(Px \rightarrow Bx)$	A	
2	(2)	$(\exists x)(Fx \wedge (Bx \rightarrow Sx))$	A	non-proof.
3	(3)	$Fa \wedge (Ba \rightarrow Sa)$	A (EE)	
4	(4)	Pa	A (CP)	
1	(5)	$Pa \rightarrow Ba$	1,UE	
1,4	(6)	Ba	4,5,MPP	
3	(7)	$Ba \rightarrow Sa$	3,$\wedge$E	
1,3,4	(8)	Sa	6,7,MPP	
1,3	(9)	$Pa \rightarrow Sa$	4,8,CP	
1,2	(10)	$Pa \rightarrow Sa$	2,3,9,EE	
1,2	(11)	$(x)(Px \rightarrow Sx)$	10,UI	

What has gone wrong? As in the last example, the arbitrary name 'a' must not occur in the conclusion C. So we mend the problem by getting rid of 'a' before discharging the assumption of line 3. Here is an amended attempt:

1	(1)	$(x)(Px \rightarrow Bx)$	A	
2	(2)	$(\exists x)(Fx \wedge (Bx \rightarrow Sx))$	A	
3	(3)	$Fa \wedge (Ba \rightarrow Sa)$	A (EE)	non-proof
4	(4)	Pa	A (CP)	
1	(5)	$Pa \rightarrow Ba$	1,UE	
1,4	(6)	Ba	4,5,MPP	
3	(7)	$Ba \rightarrow Sa$	3,$\wedge$E	
1,3,4	(8)	Sa	6,7,MPP	
1,3	(9)	$Pa \rightarrow Sa$	4,8,CP	

| 1,3 | | (10) | (x)(Px → Sx) | 9,UI |
| 1,2 | | (11) | (x)(Px → Sx) | 2,3,10,EE |

Is this one all right? No, it also is wrong, because of the restriction on the introduction of the universal quantifier, according to which the conclusion must not require any assumption in which the arbitrary name occurs. Line 10 requires the assumption of 3, in which the arbitrary name a occurs.

In fact there is no proof of the sequent. We lack the means of demonstrating that there is no proof, but a brief consideration of the original argument will satisfy us that it is incorrect. We do not know that any of the folk made smug by their benevolence are philosophers. Perhaps the study of philosophy is proof against smugness.

Exercise 4.5

Use **Simon**'s proof editor to construct a proof for each of the following sequents, using the rules of L together with UE, UI, EI, and EE.

(a)   (x)(Fx → Gx), (∃x)¬Gx ⊢ (∃x)¬Fx
(b)   (x)(Fx → Gx ∧ Hx), (∃x)Fx ⊢ (∃x)Hx
(c)   (x)(Fx ∨ Gx → Hx), (∃x)¬Hx ⊢ (∃x)¬Fx

Exercise 4.6

Use **Simon**'s proof editor to construct a proof for each of the following sequents, using the rules of L together with UE, UI, EI, and EE.

(a)   (x)(Gx → ¬Hx), (∃x)(Fx ∧ Gx) ⊢ (∃x)(Fx ∧ ¬Hx)
(b)   (x)(Hx → Gx), (∃x)(Fx ∧ ¬Gx) ⊢ (∃x)(Fx ∧ ¬Hx)
(c)   (x)(Hx → ¬Gx), (∃x)(Fx ∧ Gx) ⊢ (∃x)(Fx ∧ ¬Hx)
(d)   (x)(Gx → Hx), (∃x)(Fx ∧ Gx) ⊢ (∃x)(Fx ∧ Hx)
(e)   (x)(Gx → Fx), (∃x)(Gx ∧ Hx) ⊢ (∃x)(Fx ∧ Hx)

4.2.6 Some useful sequents with quantifiers

Here we give proofs of standard sequents that represent important formal properties of quantifiers, properties that are ordinarily taken for granted in any more advanced logical studies. The reader should study and understand the proofs, both for the results they establish, and for the technique of their construction. They will be available later for proofs of other sequents.

71   (x)(Fx → Gx) ⊢ (x)Fx → (x)Gx

1	(1)	(x)(Fx → Gx)	A
2	(2)	(x)Fx	A (CP)
2	(3)	Fa	2,UE
1	(4)	Fa → Ga	1,UE
1,2	(5)	Ga	3,4,MPP
1,2	(6)	(x)Gx	5,UI
1	(7)	(x)Fx → (x)Gx	2,6,CP

72   (x)(Fx → Gx) ⊢ (∃x)Fx → (∃x)Gx

1	(1)	(x)(Fx → Gx)	A
2	(2)	(∃x)Fx	A (CP)
3	(3)	Fa	A (EE)
1	(4)	Fa → Ga	1,UE
1,3	(5)	Ga	3,4,MPP
1,3	(6)	(∃x)Gx	5,EI
1,2	(7)	(∃x)Gx	2,3,6,EE
1	(8)	(∃x)Fx → (∃x)Gx	2,7,CP

The following pair of sequents represent what are called the *distribution properties* of the quantifiers. They are the distribution properties one would expect for connectives related as they are to ∧ and ∨. The universal quantifier distributes over ∧; the existential quantifier distributes over ∨. To say that they *distribute* over ∧ and ∨ respectively is just to say that the following two deductive equivalences hold.

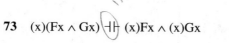

73   (x)(Fx ∧ Gx) ⊣⊢ (x)Fx ∧ (x)Gx

(⊢)   1   (1)   (x)(Fx ∧ Gx)   A   *not an assumption*
      1   (2)   Fa ∧ Ga   1,UE
      1   (3)   Fa   2,∧E
      1   (4)   (x)Fx   3,UI
      1   (5)   Ga   2,∧E
      1   (6)   (x)Gx   5,UI
      1   (7)   (x)Fx ∧ (x)Gx   4,6,∧I

(⊣)   1   (1)   (x)Fx ∧ (x)Gx   A
      1   (2)   (x)Fx   1,∧E
      1   (3)   Fa   2,UE
      1   (4)   (x)Gx   1,∧E
      1   (5)   Ga   4,UE
      1   (6)   Fa ∧ Ga   3,5,∧I
      1   (7)   (x)(Fx ∧ Gx)   6,UI

74   (∃x)(Fx ∨ Gx) ⊣⊢ (∃x)Fx ∨ (∃x)Gx

(⊢)   1   (1)   (∃x)(Fx ∨ Gx)   A
      2   (2)   Fa ∨ Ga   A (EE)   *An assumption*
      3   (3)   Fa   A (∨E)
      3   (4)   (∃x)Fx   3,EI
      3   (5)   (∃x)Fx ∨ (∃x)Gx   4,∨I
      6   (6)   Ga   A (∨E)
      6   (7)   (∃x)Gx   6,EI
      6   (8)   (∃x)Fx ∨ (∃x)Gx   7,∨I
      2   (9)   (∃x)Fx ∨ (∃x)Gx   2,3,5,6,8,∨E
      1   (10)   (∃x)Fx ∨ (∃x)Gx   1,2,9,EE

(⊣)   1   (1)   (∃x)Fx ∨ (∃x)Gx   A
      2   (2)   (∃x)Fx   A (∨E)
      3   (3)   Fa   A (EE)
      3   (4)   Fa ∨ Ga   3,∨I
      3   (5)   (∃x)(Fx ∨ Gx)   4,EI
      2   (6)   (∃x)(Fx ∨ Gx)   2,3,5,EE
      7   (7)   (∃x)Gx   A (∨E)
      8   (8)   Ga   A (EE)
      8   (9)   Fa ∨ Ga   8,∨I

8	(10)	(∃x)(Fx ∨ Gx)	9,EI
7	(11)	(∃x)(Fx ∨ Gx)	7,8,10,EE
1	(12)	(∃x)(Fx ∨ Gx)	1,2,6,7,11,∨E

As to the behaviour of the universal and existential quantifiers with ∨ and ∧ respectively, only the following single-direction provabilities hold.

75   (∃x)(Fx ∧ Gx) ⊢ (∃x)Fx ∧ (∃x)Gx

1	(1)	(∃x)(Fx ∧ Gx)	A
2	(2)	Fa ∧ Ga	A (EE)
2	(3)	Fa	2,∧E
2	(4)	(∃x)Fx	3,EI
2	(5)	Ga	2,∧E
2	(6)	(∃x)Gx	5,EI
2	(7)	(∃x)Fx ∧ (∃x)Gx	4,6,∧I
1	(8)	(∃x)Fx ∧ (∃x)Gx	1,2,7,EE

This is, intuitively, what we would expect. If two properties are instantiated in a single object, then those two properties are instantiated. We would not, however, infer from the instantiation of two properties that they are instantiated in one object. Both the property of being a cat and the property of being a dog are instantiated properties, but we not therefore suppose that there is any single object that instantiates both.

Exercise 4.7

It is instructive to see what it is about the rules that prevents the other direction. Try to prove the sequent
(∃x)Fx ∧ (∃x)Gx ⊢ (∃x)(Fx ∧ Gx).
Explain what restrictions within the rules for existential quantifiers prevent the construction of a proof.

76   (x)Fx ∨ (x)Gx ⊢ (x)(Fx ∨ Gx)

1	(1)	(x)Fx ∨ (x)Gx	A
2	(2)	(x)Fx	A (∨E)
2	(3)	Fa	2,UE
2	(4)	Fa ∨ Ga	3,∨I
2	(5)	(x)(Fx ∨ Gx)	4,UI

6	(6)	(x)Gx	A (∨E)
6	(7)	Ga	6,UE
6	(8)	Fa ∨ Ga	7,∨I
6	(9)	(x)(Fx ∨ Gx)	8,UI
1	(10)	(x)(Fx ∨ Gx)	1,2,5,6,9,∨E

Again, we do not expect the other direction to be provable. Restricting one's attention to the set of people in a room, one might expect that they are all either male or female; but one would not infer from that that either they are all male or they are all female.

Exercise 4.8

Once again, explain what restrictions within the rules for universal quantifiers prevent a proof of the sequent

(x)(Fx ∨ Gx) ⊢ (x)Fx ∨ (x)Gx,

having tried to construct a proof.

The following two sequents are the quantificational counterparts of sequent 41 (page 42) and the sequent of Ex 6 (f) (p. 47).

77   (∃x)Fx ⊣⊢ ¬(x)¬Fx

(⊢)	1	(1)	(∃x)Fx	A
	2	(2)	Fa	A (EE)
	3	(3)	(x)¬Fx	A (RAA)
	3	(4)	¬Fa	3,UE
	2,3	(5)	Fa ∧ ¬Fa	2,4,∧I
	2	(6)	¬(x)¬Fx	3,5,RAA
	1	(7)	¬(x)¬Fx	1,2,6,EE

(⊣)	1	(1)	¬(x)¬Fx	A
	2	(2)	¬(∃x)Fx	A (RAA)
	3	(3)	Fa	A (RAA)
	3	(4)	(∃x)Fx	3,EI
	2,3	(5)	(∃x)Fx ∧ ¬(Ex)Fx	2,4,∧I
	2	(6)	¬Fa	3,5,RAA
	2	(7)	(x)¬Fx	6,UI

	1,2	(8)	(x)¬Fx ∧ ¬(x)¬Fx	1,7,∧I
	1	(9)	¬¬(∃x)Fx	2,8,RAA
	1	(10)	(∃x)Fx	9,DN

78 (x)Fx ⊣⊢ ¬(∃x)¬Fx

(⊢)	1	(1)	(x)Fx	A
	2	(2)	(∃x)¬Fx	A (RAA)
	3	(3)	¬Fa	A (EE)
	1	(4)	Fa	1,UE
	1,3	(5)	Fa ∧ ¬Fa	3,4,∧I
	3	(6)	¬(x)Fx	1,5,RAA
	2	(7)	¬(x)Fx	2,3,6,EE
	1,2	(8)	(x)Fx ∧ ¬(x)Fx	1,7,∧I
	1	(9)	¬(∃x)¬Fx	2,8,RAA

(⊣)	1	(1)	¬(∃x)¬Fx	A
	2	(2)	¬Fa	A (RAA)
	2	(3)	(∃x)¬Fx	2,EI
	1,2	(4)	(∃x)¬Fx ∧ ¬(∃x)¬Fx	1,3,∧I
	1	(5)	¬¬Fa	2,4,RAA
	1	(6)	Fa	5,DN
	1	(7)	(x)Fx	6,UI

The deductively equivalent sentences of the following sequent are called *alphabetic variants*. Natural languages have small numbers of pronouns (he, she, it, they, and so) which have specific roles. A quantificational language typically has a denumerably infinite supply of individual variables, which are the formal counterparts of pronouns. Though there is a need for the infinite supply, since sentences can be of any finite length, there is nothing distinctive about any of these formal pronouns. In the simplest cases, those of the next two sequents, any pair of individual variables are interchangeable.

79 (x)Fx ⊣⊢ (y)Fy

(⊢)	1	(1)	(x)Fx	A
	1	(2)	Fa	1,UE
	1	(3)	(y)Fy	2,UI

(⊣)	1	(1)	(y)Fy	A
	1	(2)	Fa	1,UE
	1	(3)	(x)Fx	2,UI

80   (∃x)Fx ⊣⊢ (∃y)Fy

(⊢)	1	(1)	(∃x)Fx	A
	2	(2)	Fa	A (EE)
	2	(3)	(∃y)Fy	2,EI
	1	(4)	(∃y)Fy	1,2,3,EE

(⊣)	1	(1)	(∃y)Fy	A
	2	(2)	Fa	A (EE)
	2	(3)	(∃x)Fx	2,EI
	1	(4)	(∃x)Fx	1,2,3,EE

The following sequent can be understood as the quantificational counterpart of **40** (p. 42). Such a sentence as the lefthand sentence of the equivalence is sometimes referred to as asserting a *formal implication*. The equivalence then 'says' that F formally implies G if and only if nothing has the former property that does not have the latter.

81   (x)(Fx → Gx) ⊣⊢ ¬(∃x)(Fx ∧ ¬Gx)

(⊢)	1	(1)	(x)(Fx → Gx)	A
	2	(2)	(∃x)(Fx ∧ ¬Gx)	A (RAA)
	3	(3)	Fa ∧ ¬Ga	A (EE)
	3	(4)	¬(Fa → Ga)	3,SI(S) 3.5(g)
	1	(5)	Fa → Ga	1,UE
	1,3	(6)	(Fa → Ga) ∧ ¬(Fa → Ga)	4,5,∧I
	3	(7)	¬(x)(Fx → Gx)	1,6,RAA
	2	(8)	¬(x)(Fx → Gx)	2,3,7,EE
	1,2	(9)	(x)(Fx → Gx) ∧ ¬(x)(Fx → Gx)	1,8,∧I
	1	(10)	¬(∃x)(Fx ∧ ¬Gx)	2,9,RAA

(⊣)	1	(1)	¬(∃x)(Fx ∧ ¬Gx)	A
	2	(2)	¬(Fa → Ga)	A (RAA)
	2	(3)	Fa ∧ ¬Ga	2,SI(S) 3.5(g)
	2	(4)	(∃x)(Fx ∧ ¬Gx)	3,EI
	1,2	(5)	(∃x)(Fx ∧ ¬Gx) ∧ ¬(∃x)(Fx ∧ ¬Gx)	1,4,∧I

1	(6)	¬¬(Fa → Ga)	2,5,RAA
1	(7)	Fa → Ga	6,DN
1	(8)	(x)(Fx → Gx)	7,UI

The next equivalence is the quantificational counterpart of the propositional equivalence

$$(P \to R) \land (Q \to R) \dashv\vdash (P \lor Q) \to R$$

the left-to-right direction of which was proved in Exercise 2.3 (f) (p. 35). The quantificational equivalence as given here is restricted to those cases in which there are no occurrences of the individual variable in the consequent of the conditional. The restriction is an important one in weighing alternative formal representations of English conditionals. Both alternatives are available as representations of those English conditionals in which there is no pronominal reference from the *then*-clause to a quantified noun phrase in the *if*-clause. Thus for example, the sentence

> If any member votes I will be surprised

can be represented by either

$$(x)(Fx \to P)$$

or by

$$(\exists x)Fx \to P$$

but the sentence

> If any member votes, she will be rewarded

can be represented by

$$(x)(Fx \to Gx)$$

which has no correspondingly equivalent form using the existential quantifier. We will revert to the subject before the close of this chapter.

82 (x)(Fx → P) ⊣⊢ (∃x)Fx → P

(⊢)	1	(1)	$(x)(Fx \to P)$	A
	2	(2)	$(\exists x)Fx$	A (CP)
	3	(3)	Fa	A (EE)
	1	(4)	$Fa \to P$	1,UE
	1,3	(5)	P	3,4,MPP
	1,2	(6)	P	2,3,5,EE
	1	(7)	$(\exists x)Fx \to P$	2,6,CP

(⊣)	1	(1)	$(\exists x)Fx \to P$	A
	2	(2)	Fa	A (CP)
	2	(3)	$(\exists x)Fx$	2,EI
	1,2	(4)	P	1,3,MPP
	1	(5)	$Fa \to P$	2,4,CP
	1	(6)	$(x)(Fx \to P)$	5,UI

83 $(\exists x)(P \to Fx) \dashv\vdash P \to (\exists x)Fx$

(⊢)	1	(1)	$(\exists x)(P \to Fx)$	A
	2	(2)	P	A (CP)
	3	(3)	$P \to Fa$	A (EE)
	2,3	(4)	Fa	2,3,MPP
	2,3	(5)	$(\exists x)Fx$	4,EI
	1,2	(6)	$(\exists x)Fx$	1,3,5,EE
	1	(7)	$P \to (\exists x)Fx$	2,6,CP

(⊣)	1	(1)	$P \to (\exists x)Fx$	A
		(2)	$P \vee \neg P$	TI 51
	3	(3)	P	A ($\vee$E)
	1,3	(4)	$(\exists x)Fx$	1,3,MPP
	5	(5)	Fa	A (EE)
	5	(6)	$P \to Fa$	5,SI(S)57
	5	(7)	$(\exists x)(P \to Fx)$	6,EI
	1,3	(8)	$(\exists x)(P \to Fx)$	4,5,7,EE
	9	(9)	$\neg P$	A ($\vee$E)
	9	(10)	$P \to Fa$	9,SI(S)58
	9	(11)	$(\exists x)(P \to Fx)$	10,EI
	1	(12)	$(\exists x)(P \to Fx)$	2,3,8,9,11,$\vee$E

Exercise 4.9

Use **Simon**'s proof editor to construct a proof for each of the following sequents.

(a)   (x)(Fx → Gx) ⊢ (x)¬Gx → (x)¬Fx
(b)   (x)(Fx → Gx) ⊢ (∃x)¬Gx → (∃x)¬Fx
(c)   (∃x)¬Fx ⊣⊢ ¬(x)Fx
(d)   (x)¬Fx ⊣⊢ ¬(∃x)Fx
(e)   (x)(Fx → ¬Gx) ⊣⊢ ¬(∃x)(Fx ∧ Gx)
(f)   (x)(Fx ↔ Gx) ⊣⊢ (x)(Fx → Gx) ∧ (x)(Gx → Fx)
(g)   (x)(Fx ↔ Gx) ⊢ (x)Fx ↔ (x)Gx
(h)   (x)(Fx ↔ Gx) ⊢ (∃x)Fx ↔ (∃x)Gx

Exercise 4.10

Use **Simon**'s proof editor to construct proofs for the following deductive equivalences.

(a)   (x)(P → Fx) ⊣⊢ P → (x)Fx
(b)   (x)(P ∧ Fx) ⊣⊢ P ∧ (x)Fx
(c)   (∃x)(P ∧ Fx) ⊣⊢ P ∧ (∃x)Fx
(d)   (x)(P ∨ Fx) ⊣⊢ P ∨ (x)Fx
(e)   (∃x)(P ∨ Fx) ⊣⊢ P ∨ (∃x)Fx
(f)   (∃x)Fx → P ⊣⊢ (x)(Fx → P)

Exercise 4.11

In each of the following sets of quantificational sentences, at least one of the sentences is provable from the rest. In each case use **Simon**'s Sequent Editor to construct as many provable sequents as you can, and give proofs for them using the Proof Editor.

(a)   Gm
      Fm
      (x)(Fx → Gx)

(b)   (y)(Hy → ¬Fy)
      (x)(Fx → Gx)
      (y)(Hy → ¬Gy)
      (x)(Fx → ¬Hx)

(c) $(\exists x)(Fx \wedge Gx)$
$(\exists x)(Fx \wedge \neg Hx)$
$(x)(Gx \to \neg Hx)$

(d) $(x)(Gx \leftrightarrow Hx)$
$(x)(Fx \vee Gx \to Hx \wedge Jx)$
$(x)(Hx \vee Kx \to Gx \wedge Lx)$

(e) $(x)(Gx \to Fx)$
$(x)(Fx \to Gx)$
$(\exists y)(Fy \wedge \neg Gy)$
$(\exists y)(Gy \wedge \neg Fy)$
$(x)(Fx \to (Gx \to Fx))$

(f) $(x)(Jx \vee Kx \to ((y)((Fy \to Gy) \vee (Gy \to Fy)) \to Fx \wedge Kx))$
$(x)(Fx \leftrightarrow Jx)$
$(x)(Gx \leftrightarrow Kx)$
$(x)(Fx \vee Gx \to ((y)(\neg Fy \to (Fy \to Gy)) \to Jx \wedge Kx))$

(g) $(x)(Fx \to Gx)$
$(x)(Fx \to Hx)$
$(x)(Fx \to (Gx \to Hx))$

4.2.7 Nested quantifiers

All of the sentences occurring in proofs thus far have had only 1-place occurrences of predicate letters. However, in general the language of quantificational logic permits n-place occurrences of predicate letters for any n. Sentences having n-place occurrences of predicate letters for n greater than 1 will in general have successions of quantifiers, and it is important to establish in which cases the order of the quantifiers is inferentially significant, and when it is not. In general, multiple universal quantifiers or multiple existential quantifiers can occur equivalently, as the following illustrate.

84 $(x)(y)Fxy \dashv\vdash (y)(x)Fxy$

($\vdash$)	1	(1)	$(x)(y)Fxy$	A
	1	(2)	$(y)Fay$	1,UE
	1	(3)	Fab	2,UE

1	(4)	(x)Fxb	3,UI
1	(5)	(y)(x)Fxy	4,UI

(⊣)	1	(1)	(y)(x)Fxy	A
	1	(2)	(x)Fxb	1,UE
	1	(3)	Fab	2,UE
	1	(4)	(y)Fay	3,UI
	1	(5)	(x)(y)Fxy	4,UI

85   (∃x)(∃y)Fxy ⊣⊢ (∃y)(∃x)Fxy

(⊢)	1	(1)	(∃x)(∃y)Fxy	A
	2	(2)	(∃y)Fay	A (EE)
	3	(3)	Fab	A (EE)
	3	(4)	(∃x)Fxb	3,EI
	3	(5)	(∃y)(∃x)Fxy	4,EI
	2	(6)	(∃y)(∃x)Fxy	2,3,5,EE
	1	(7)	(∃y)(∃x)Fxy	1,2,6,EE

(⊣)	1	(1)	(∃y)(∃x)Fxy	A
	2	(2)	(∃x)Fxb	A (EE)
	3	(3)	Fab	A (EE)
	3	(4)	(∃y)Fay	3,EI
	3	(5)	(∃x)(∃y)Fxy	4,EI
	2	(6)	(∃x)(∃y)Fxy	2,3,5,EE
	1	(7)	(∃x)(∃y)Fxy	1,2,6,EE

Exercise 4.12

Although homogeneous successions of universal or existential quantifiers are order-insensitive, not all quantifiers of natural language have this property. It has been known since the middle ages that if we introduce a quantifier (μx) having the reading 'it is true as regards most individuals x that . . .', then homogeneous successions of such quantifiers will not be order-insensitive. Demonstrate the inequivalence of (μx)(μy)Fxy and (μy)(μx)Fxy by constructing a counterexample using two groups of five individuals. That is, describe a case in which for some relation R, it is true as regards most of group A that they are in the relation to most of group B, but false of most of group B, that most of group A are in the relation R to them.

Mixed successions of universal and existential quantifiers are order-sensitive. The converse of the following sequent is not provable.

86   (∃x)(y)Fxy ⊢ (y)(∃x)Fxy

1	(1)	(∃x)(y)Fxy	A
2	(2)	(y)Fay	A (EE)
2	(3)	Fab	2,UE
2	(4)	(∃x)Fxb	3,EI
2	(5)	(y)(∃x)Fxy	4,UI
1	(6)	(y)(∃x)Fxy	1,2,5,EE

Exercise 4.13

Intuitively, we should not expect to be able to prove the converse of the previously proved sequent. Certainly if there is some bicycle that every girl rides, then every girl rides a bicycle, but we do not infer from the supposition that every girl rides a bike that they all ride the same one. Try to construct a proof of the converse of the sequent, noting what restrictions within the quantificational rules block the proof.

4.2.8 Some examples of arguments

In his famous Cambridge lectures, Augustus DeMorgan, alluded to a class of evidently correct arguments for which the received traditional logic would not permit a proof. His example was

> All horses are animals; therefore all heads of horses are heads of animals.

In fact, Aristotle, the ultimate source of the logical tradition against which DeMorgan was contending, was not oblivious of the the case, and had tried to make theoretical provision for such arguments. Twentieth-century quantificational logic does, however, permit relational arguments of this sort to be represented and the system of quantificational rules does permit the construction of a proof. With Fx (x is a horse), Gx (x is an

animal), and Hxy (x is the tail of y), the argument can be represented as
the sequent

87 (x)(Fx → Gx) ⊢ (x)((∃y)(Fy ∧ Hxy) → (∃y)(Gy ∧ Hxy))

1	(1)	(x)(Fx → Gx)	A
2	(2)	(∃y)(Fy ∧ Hay)	A(CP)
3	(3)	Fb ∧ Hab	A(EE)
3	(4)	Fb	3,∧E
1	(5)	Fb → Gb	1,UE
1,3	(6)	Gb	4,5,MPP
3	(7)	Hab	3,∧E
1,3	(8)	Gb ∧ Hab	6,7,∧I
1,3	(9)	(∃y)(Gy ∧ Hay)	8,EI
1,2	(10)	(∃y)(Gy ∧ Hay)	2,3,9,EE
1	(11)	(∃y)(Fy ∧ Hay) → (∃y)(Gy ∧ Hay)	2,10,CP
1	(12)	(x)((∃y)(Fy ∧ Hxy) → (∃y)(Gy ∧ Hxy))	11,UI.

As a second example, consider the argument

> Some boys like all games. No boys like dull things. Therefore, no
> games are dull things.

With Fx (x is a boy), Gx (x is a game), Jx (x is dull), Hxy (x likes y), one
representing sequent would be

88 (∃x)(Fx ∧ (y)(Gy → Hxy)), (x)(y)(Fx → (Jy → ¬Hxy))
⊢ (x)(Gx → ¬Jx)

1	(1)	(∃x)(Fx ∧ (y)(Gy → Hxy))	A
2	(2)	(x)(y)(Fx → (Jy → ¬Hxy))	A
3	(3)	Ga	A (CP)
4	(4)	Fb ∧ (y)(Gy → Hby)	A (EE)
4	(5)	(y)(Gy → Hby)	4,∧E
4	(6)	Ga → Hba	5,UE
3,4	(7)	Hba	3,6,MPP
2	(8)	(y)(Fb → (Jy → ¬Hby))	2,UE
2	(9)	Fb → (Ja → ¬Hba)	8,UE
4	(10)	Fb	4,∧E
2,4	(11)	Ja → ¬Hba	9,10,MPP

3,4	(12)	¬¬Hba	7,DN
2,3,4	(13)	¬Ja	11,12,MTT
1,2,3	(14)	¬Ja	1,4,13,EE
1,2	(15)	Ga → ¬Ja	3,14,CP
1,2	(16)	(x)(Gx → ¬Jx)	15,UI

Notice that in the construction of a proof, the EE rule might allow us a good deal of latitude in the matter of where we discharge the assumption. Here we could have applied EE after line 11; alternatively we could have waited and applied it as the last step of the proof. In general we apply EE wherever clarity and convenience dictate, provided that the conditions governing required assumptions have been met. In this application, the sentence ¬Ja at line 13, since it contains fewer symbols, requires less rewriting than any of the alternatives.

As a third example, consider the argument

> Some botanists are eccentrics. Some botanists do not like any eccentric; therefore some botanists are not liked by all botanists.

Using Fx (x is a botanist), Gx (x is an eccentric), Hxy (x likes y), the argument can be represented by the sequent:

89 (∃x)(Fx ∧ Gx), (∃x)(Fx ∧ (y)(Gy → ¬Hxy))
⊢ (∃x)(∃y)((Fx ∧ Fy) ∧ ¬Hyx)

1	(1)	(∃x)(Fx ∧ Gx)	A
2	(2)	(∃x)(Fx ∧ (y)(Gy → ¬Hxy))	A
3	(3)	Fa ∧ Ga	A (EE)
3	(4)	Fa	3,∧E
3	(5)	Ga	3,∧E
6	(6)	Fb ∧ (y)(Gy → ¬Hby)	A (EE)
6	(7)	Fb	6,∧E
6	(8)	(y)(Gy → ¬Hby)	6,∧E
6	(9)	Ga → ¬Hba	8,UE
3,6	(10)	¬Hba	5,9,MPP
3,6	(11)	(Fa ∧ Fb)	4,7,∧I
3,6	(12)	(Fa ∧ Fb) ∧ ¬Hba	10,11,∧I
3,6	(13)	(∃y)((Fa ∧ Fy) ∧ ¬Hya)	12,EI

3,6	(14)	$(\exists x)(\exists y)((Fx \wedge Fy) \wedge \neg Hyx)$	13,EI
2,3	(15)	$(\exists x)(\exists y)((Fx \wedge Fy) \wedge \neg Hyx)$	2,6,14,EE
1,2	(16)	$(\exists x)(\exists y)((Fx \wedge Fy) \wedge \neg Hyx)$	1,3,15,EE

A final example involves a 3-place predicate:

If anyone speaks to anyone, then someone introduces them. No one introduces anyone to anyone unless he knows them both. Everyone speaks to William. Therefore, everyone is introduced to William by someone who knows him.

Using Fxy (x speaks to y), Gxy (x knows y), Hxyz (x introduces y to z), m (William), and (permissibly) suppressing the common property of being a person, the argument can be represented by the sequent

90 $(x)(y)(Fxy \rightarrow (\exists z)Hzxy), (x)(y)(z)(Hxyz \rightarrow Gxy \wedge Gxz), (x)Fxm$
$\vdash (x)(\exists z)(Hzxm \wedge Gzm)$

1	(1)	$(x)(y)(Fxy \rightarrow (\exists z)Hzxy)$	A
2	(2)	$(x)(y)(z)(Hxyz \rightarrow Gxy \wedge Gxz)$	A
3	(3)	$(x)Fxm$	A
3	(4)	Fam	3,UE
1	(5)	$(y)(Fay \rightarrow (\exists z)(Hzay))$	1,UE
1	(6)	$Fam \rightarrow (\exists z)Hzam$	5,UE
1,3	(7)	$(\exists z)Hzam$	4,6,MPP
8	(8)	$Hbam$	A (EE)
2	(9)	$(y)(z)(Hbyz \rightarrow Gby \wedge Gbz)$	2,UE
2	(10)	$(z)(Hbaz \rightarrow Gba \wedge Gbz)$	9,UE
2	(11)	$Hbam \rightarrow Gba \wedge Gbm$	10,UE
2,8	(12)	$Gba \wedge Gbm$	8,11,MPP
2,8	(13)	Gbm	12,$\wedge$E
2,8	(14)	$Hbam \wedge Gbm$	8,13,$\wedge$I
2,8	(15)	$(\exists z)(Hzam \wedge Gzm)$	14,EI
1,2,3	(16)	$(\exists z)(Hzam \wedge Gzm)$	7,8,15,EE
1,2,3	(17)	$(x)(\exists z)(Hzxm \wedge Gzm)$	16, UI

4.2.9 On representing English conditionals

$$\boxed{A}$$

> "If every…"
>
> "If all…"   become: $(x)(\ldots) \rightarrow (\ldots)$
>
> "If no…"

$$\boxed{B}$$

> "If any…"
>
> Look in the *then*-clause for a pronoun (*he, it, her, him, she, them, they*) referring to the quantified noun phrase of the *if*-clause. If it is PRESENT then proceed to $\boxed{\textbf{PRESENT}}$; if it is ABSENT then proceed to $\boxed{\textbf{ABSENT}}$.
>
> "If some…"
>
> > $$\boxed{\textbf{PRESENT}}$$
> >
> > "If any…"
> >   become: $(x)(\ldots \rightarrow \ldots)$
> > "If some…"
>
> > $$\boxed{\textbf{ABSENT}}$$
> >
> > "If any…"
> >   become either: $(x)(\ldots \rightarrow \ldots)$
> >   or, if you wish: $(\exists x)(\ldots) \rightarrow (\ldots)$
> > "If some…"

The foregoing is a flow-chart-of-thumb for representing conditionals in which a quantified noun phrase (a noun phrase with *all, any, every, no, some*) follows the "if".There are two kinds of case: $\boxed{A}$ those in which "if" is followed by "all" or "every", and $\boxed{B}$ those in which "if" is followed by "any" or "some". English has its little ways with this latter vocabulary, and the meanings of conditional constructions are sensitive to the vocabulary used. In general, *all, every, no* are read as short-scope quantifiers, but *any* is read as a long-scope quantifier, though these readings can be contextually precluded. (Consider the sentence *If (just) anyone is allowed to join, then I don't want to.*) The parentheses of the schemata provided are intended to indicate the scope of the quantifier in each case. As we have already seen in an earlier discussion of *or* in antecedents,the problems are not specific to quantifiers. They are problems of grammatical reference.

Exercise 4.14

Consulting the flowchart for guidance as required, and using **Simon**'s formulation editor, represent each of the following sentences in quantificational language, using the predicate letters provided. (The individual variables are given only to indicate the number of places of the predicate letter, not to dictate the use of those variables.)

(a) If anyone cheats, then someone is punished (Fx, Gx, Hx)
(b) If someone cheats, then she is punished (Fx, Gx, Hx)
(c) If anyone cheats, then anyone who knows her is ashamed of her. (Fx, Gx, Hxy, Jxy)
(d) If something is destroyed, then someone is gratified (Fx, Gx, Hx)
(e) If anything is destroyed, then someone pays for it (Fx, Gx, Hxy)
(f) If anything is destroyed, then someone pays someone something (Fx, Gx, Hxyz)

Exercise 4.15

Use **Simon**'s proof editor to construct a proof for each of the following sequents:

(a) $(x)(y)(z)Fxyz \vdash (z)(y)(x)Fxyz$
(b) $(x)(\exists y)(z)Fxyz \vdash (x)(z)(\exists y)Fxyz$
(c) $(\exists x)(\exists y)(z)Fxyz \vdash (z)(\exists y)(\exists x)Fxyz$

Exercise 4.16

Following the instructions given in Exercise 4.3, represent each of the following arguments as a sequent, and construct a proof for the sequent.

(a) If it rains, no birds are happy. If it snows, some birds are happy. Therefore if it rains, then it does not snow. (P (It rains), Q (It snows))

(b) All camels like a gentle driver. Some camels do not like Mohammed. Mohammed is a driver. Therefore, Mohammed is not gentle.

(c) All camels are highly strung. Some drivers like no highly strung animals. Camels are animals. Therefore, some drivers do not like any camels.

(d) Some dogs like William. All boys like any dog. William is a boy. Therefore there is something that both likes and is liked by William.

(e) A whale is a mammal. Some fish are whales. All fish have tails. Therefore, some fishes' tails are mammals' tails.

5

Quantificational Logic 2

versus Propositional logic

5.1 The formal system L_q

We have by now acquired a sufficient informal acquaintance with quantificational procedures that we will benefit from a proper and orderly definition of a system, along the lines of the definition of the system L given in Chapter 2. Recall that a formal system is a pair consisting of a language and a set of rules. Recall also that a language is a triple consisting of a set of atoms, a set of constants, and an inductively defined set of wffs. For the simpler language of the system L, the set At of atomic sentences was just the set of sentence letters, each consisting of (what we defined as) exactly one symbol. Because the atomic sentences of propositional logic cannot be further broken down into subatomic items, propositional logic is sometimes referred to as *the logic of unanalysed propositions.* By contrast, the smallest sentences considered in quantificational logic comprise two distinct kinds of subatomic component: a predicate letter and a set of names (in some systems, a predicate letter and a set of names and individual variables). Moreover, the distinction between names and individual variables raises issues that warrant a distinction between the general notion of atom and the more specific notion of an atomic wff. For these reasons, quantificational logic is sometimes referred to as *the logic of analysed propositions.* Before we can give an account of the atoms, we must give an account of the subatomic symbols out of which atoms can be constructed.

As we have seen, the system of quantificational logic we are introducing extends rather than replaces the system of propositional logic with which we began. Our description of the system therefore considers the elements of a language one by one, enriching each by the addition of required items. All of the previous items remain. We begin with an account of the new symbols, of which there are five categories.

1. Proper names A proper name is any of a set of marks:

$$l, m, n, l', l'' \ldots$$

2. Arbitrary names An arbitrary name is any of a denumerably infinite set of marks:

$$a, b, c, a', a'' \ldots$$

3. Individual variables An individual variable is any of a denumerably infinite set of marks:

$$x, y, z, x', x'' \ldots$$

4. Predicate letters A predicate letter is any of a denumerably infinite set of marks:

$$F, G, H, F', F'' \ldots$$

These are understood to include the sentence letters of propositional language.

5. Reverse-E Reverse-E is the mark:

$$\exists$$

A *term* is any mark from category 1 or 2, that is, a proper or an arbitrary name. A *nominal* is any mark that is either a term or an individual variable. A *symbol* is either a bracket or a term or an individual variable or a predicate letter (including sentence letters, that is, propositional atoms) or reverse-E. A *formula* is any finite sequence of symbols.

5.1.1 Atomic wffs

As for propositional language, our description requires a stock of *metalogical variables*. A, B, C, . . . are metalogical variables ranging over formulae; 'P' is a metalogical variable ranging over predicate letters; t_1, t_2, t_3, . . . are metalogical variables ranging over terms; v_1, v_2, v_3, . . . are metalogical variables ranging over individual variables; l_1, l_2, l_3, . . . are metalogical variables ranging over *nominals*, that is, over terms and individual variables.

Let t_1, . . . , t_n be n terms (not necessarily distinct), for n ≥ 0, and P any predicate letter. Then

$$Pt_1...t_n$$

is an *atomic sentence*. Thus, an atomic sentence is any whole formula consisting of a predicate letter followed by any finite number of occurrences of terms. For example,

F, Fm, Fmn, Gamn, Hmmmmmmmmmmmmm   Atomic sentence

are all atomic sentences. However,

Fx, Fmx, Gayn, Hmmmmmmmmmmmmz   not an atomic sentence

are not, because 'x', 'y', and 'z' are not terms. As we have seen, predicate letters followed by no terms at all count as atomic sentences. However, our convention will continue to be to use the sentence letters of propositional logic in this role.

5.1.2 The set Φ (Φq) of wffs

(a) Every atomic sentence is a wff.

(b) If A is a wff, then ¬A is a wff.

(c) If A is a wff and B is a wff, then (A → B) is a wff.

(d) If A is a wff and B is a wff, then (A ∧ B) is a wff.

(e) If A is a wff and B is a wff, then (A ∨ B) is a wff.

(f) Let A(t) be a wff containing the term t, and let v be some individual variable not occurring in A(t); let A(v) be a formula obtained from A(t) by replacing at least one occurrence of t by v. Then (v)A(v) is a wff.

(g) Let A(t) be a wff containing the term t, and let v be some individual variable not occurring in A(t); let A(v) be a formula obtained from A(t) by replacing at least one occurrence of t by v. Then (∃v)A(v) is a wff.

(h) No formula is a wff if it not a wff by one of the preceding clauses.

The language of our new system is an enrichment of the language of our old. The atomic wffs are just the atomic sentences, which includes the At of propositional Φ. The propositional connectives →, ¬, ∧, ∨ all remain in use. We have included a clause for ↔ as a matter of convenience. Clauses (f) and (g) in effect introduce two new unary connectives: the universal and existential quantifiers. These last clauses deserve a little attention. To show that

$$(x)(Fx \rightarrow Gx)$$

is a wff, note that

$$Fa \rightarrow Ga \qquad (A(a))$$

is a wff (since Fa and Ga are atomic sentences). Moreover,

$$Fx \rightarrow Gx \qquad (A(x))$$

results from A(a) by the substitution of x for both occurrences of a. Therefore,

$$(x)(Fx \rightarrow Gx) \qquad (x)A(x)$$

is a wff. And by similar reasoning, (x)(Fx → Ga), and (y)(Fa → Gy) are wffs. Notice that *the variable must replace at least one occurrence* of the term.

$$(x)(Fa \rightarrow Ga)$$

is not a wff, because it cannot be got by the procedure given in (f). In particular, no occurrence of 'a' has been replaced by an occurrence of 'x' Again notice that *the variable being substituted must not occur in the wff A(t)*. Let A(t) be (∃x)(Fx → Ga). The formula obtained by substituting 'x' for 'a' and adding a quantifier in 'x':

$$(x)(x)(Fx \rightarrow Gx) \text{ and } (x)(\exists x)(Fx \rightarrow Gx)$$

are not well-formed, although

$$(y)(x)(Fx \rightarrow Gy) \text{ and } (y)(\exists x)(Fx \rightarrow Gy)$$

are. And again, the formula (x)(Fx → (x)Gx) is not well-formed. To have obtained a wff from (Fa → (x)Gx) or (Fm → (x)Gx), we would have had to replace the occurrence of 'a' ('m') with an occurrence of an individual variable such as 'y', having no occurrences already in the wff. This would produce the formula

$$(y)(Fy \rightarrow (x)Gx)$$

which is well-formed.

5.1.3 Propositional functions

A formula such as Fx or (x)Fxy is not well-formed by our definition, but it clearly is something significant. Historically such constructions have been called *propositional functions* because when the individual variable x (or in the second case y) takes some individual as its value, what was called *a proposition* results, that is, an item having a truth-value. So a propositional function (in one variable) is a function from a set of individuals to a set of propositions. Some English sentences containing pronouns could be regarded as propositional functions.

Consider the sentence

>He is prodigiously clever.

If George (who *is* prodigiously clever) is made the referent of the 'he', then the sentence takes George (considered as its input) to a true proposition. But the sentence takes Clovis (who is rather stupid) to a false proposition.

Here we require a purely syntactic definition of *propositional function*. Accordingly we shall say that a formula A is a *propositional function in the variables* $v_1, \ldots, v_n$, for $n \geq 0$, if $(v_1)\ldots(v_n)A$ is a wff. Thus Fx is a (an atomic) propositional function in one variable, x, because (x)Fx is a wff; Fxyy is a propositional variable in two variables, x and y, because (x)(y)Fxyy is a wff; and so on. An *atomic propositional function* is a formula consisting of a predicate letter followed by n nominals. In general, the formula resulting from the dropping of n quantifiers from a wff is a propositional function in n variables. Since we also allow the case in which $n = 0$, every wff is a propositional function in 0 variables. Thus the expression *propositional function* as we shall use it is broader than the expression *well-formed formula*. In clause (g) of the definition of Φ (as we shall refer to what would more properly be called Φ_q) A(v) is a propositional function in the variable v.

The quantifiers can be syntactically defined as follows: a left bracket followed by an individual variable, followed by a right bracket is called a *universal quantifier*; a left bracket followed by a reverse-E, followed by an individual variable, followed by a right bracket is called an *existential quantifier*. A *quantifier* is either a universal quantifier or an existential quantifier. The quantificational notion of the *scope* of a connective is essentially the propositional notion of scope adapted to the richer quantificational setting.

>The scope of an occurrence of a connective in a propositional function is the shortest propositional function in which it occurs.

Thus the scope of the sole occurrence of '$\rightarrow$' in the propositional function, $(y)(Fy \rightarrow (x)Gx)$ is the propositional function in y, $(Fy \rightarrow (x)Gx)$. By the same token, the scope of an occurrence of a quantifier in a propositional function is the shortest propositional function in which it occurs. In the last mentioned wff, the scope of the sole occurrence of (y) is the whole wff; the scope of the occurrence of (x) is (x)Gx.

By reason of the clauses (g) and (h) in the definition of Φ, we must take note of the following:

(1)   the scope of an occurrence of a quantifier in a propositional function must contain at least two occurrences of the individual variable of the quantifier.

(2)   the scope of an occurrence of a quantifier in a propositional function must not contain any other quantifier having the same individual variable.

The formula (x)(Fa → Ga) fails to satisfy the former; (x)(x)(Fx → Gx) fails to satisfy the latter requirement. There is a certain latitude in how a quantificational language is formulated, and in some systems both of those formulae would be tolerated. However, a formal system has two elements: a language and a set of rules. The manner of formulation of the one affects that of the other. Here we strive for a particular simplicity in the statement of the quantificational rules that cannot be achieved in the presence of formulae such as the ones that conditions (1) and (2) preclude. Since, in addition, such formulae give the language no new expressive power, we lose nothing in excluding them.

As in the propositional language of Chapter 2, we omit outermost brackets, and others where the scope conventions for propositional connectives permit no ambiguity.

Finally, if $A_1, \ldots, A_n$, are wffs of Φ (Φ_q) containing no occurrences of arbitrary names[1] and B is a wff of Φ, (Φ_q),

$$A_1, \ldots, A_n \vdash B$$

---

1. The reader may have noticed the absence of arbitrary names in the premises of the sequents proved so far. The reason for their exclusion in the definition of *sequent*, will become evident in the sequel.

is a sequent of the system L_q.

page 45

5.2 The Rules of L_q

All ten rules of L are rules of L_q, where their use is extended to quantificational substitution instances of the wffs of Φ. Thus, for example, if a proof contains as a line, the wff (x)Fx, the proof can be extended to contain as a line the wff (x)Fx $\vee$ A for any wff A, and with the same assumption-requirements as for propositional logic. In addition to those rules, we adopt introduction and elimination rules for universal and existential quantifiers. These can now be formulated more carefully in view of the more precise account that we have now given of the language Φ_q.

The rules are most easily presented in pairs, accordingly as their use must satisfy one or the other of two *substitutional requirements*, the conditions governing the replacement of variables by names, and the replacement of names by variables in applications of the rules. UE and EI share one substitutional requirement; UI and EE share the other. In the latter pair there are additional restrictions expressed in *side conditions*.

5.2.1 UE and EI

Shared substitutional requirement

A(v) is a propositional function in v, and t is a term. A(t) is the result of replacing all and only occurrences of v in A(v) by t.

UE   A proof of (v)A(v) from Σ can be extended to a proof of A(t) from Σ.

EI   A proof of A(t) from Σ be extended to a proof of ($\exists$v)A(v) from Σ.

Notes:

Bear in mind that the substitutional requirement does not describe an operation that is performed in the course of constructing the proof. It just describes an asymmetric relationship between the input sentence and the output sentence of the rules. In each case, the pair must satisfy the condition that one of them can be got from the other by a substitution of occurrences of a term for occurrences of an individual variable. In the case of UE, the direction of that operation is the same as the direction of the proof. That is, the output wff must be capable of being produced by the substitution of occurrences of a term for occurrences of a variable in part of the input wff. But in the case of EI, the requirement applies in the opposite direction: the *input* wff must be capable of being produced from part of the output wff by that operation. Consider a pair of propositional functions that satisfy the substitutional requirement: A(m), Fmm and A(x), Fxm. Fmm can be produced from Fxm by replacement of all and only occurrences of the variable x by the term m. Accordingly UE lets us write down the line Fmm, given the line (x)Fxm. Intuitively, imagine the inference from 'Everyone loves Fred' to 'Fred loves Fred'. EI, by contrast, permits us to write down the line (∃x)Fxm, given the line Fmm. Intuitively, imagine the inference from 'Fred loves Fred' to 'Someone loves Fred'. In fact, the narcissistic Fred is a good mascot for this rule. Remember that if Fred loves Fred, then all of the following hold: Someone loves Fred; Fred loves someone; someone loves someone.

5.2.2 UI and EE

Shared substitutional requirement

A(e) is a wff containing an occurrence of the arbitrary name e, and v a variable not occurring in A(e). A(v) is the propositional function which results from the replacement of all and only occurrences of e in A(e) by v.

UI   A proof of A(e) from Σ can be extended to a proof of (v)A(v) from Σ, provided that:

Side condition: e occurs in no assumptions required by A(e).

EE   A proof that includes a proof of (∃v)A(v) from Σ, and that includes a proof of wff **C** from Δ, A(e) can be extended to a proof of **C** from Σ,Δ, provided that:

Side condition: e does not occur in **C** or in any assumption required by **C** except A(e).

Notes:

In this case, the substitutional requirement mentions the replacement of all occurrences of an arbitrary name by occurrences of a variable. In the first instance, we can remark, it should be no surprise that EE and UI should be linked in this way.

Notice first that the side-conditions of the two rules are also connected. Remember that on our intuitive justification for EE, we must demonstrate that the conclusion follows *whichever* individual is the basis for the existentially quantified premise. We must in effect therefore prove a universally quantified sentence to be justified in drawing an EE conclusion. This can be made precise within the terms our account of the rule. Imagine constructing a proof in which EE is to be used. There is a line, say, (∃x)Fx; you have made the assumption Fa; you then construct a proof of C from the assumption Fa together with some other assumptions. The last line of that proof is C.

Quaere: what conditions must C satisfy in order for it to be written down again as an EE-justified line? The conditions are exactly the conditions that would have to be met were you to write down Fa → C as a CP-justified line, and then (x)(Fx → C) as a UI-justified line. For EE, C is permitted to have A(e) as a required assumption in which e appears. In the contemplated CP step, that assumption would be dropped, leaving no assumptions with occurrences of 'e' to thwart a UI step. That is to say, the conditions that C must satisfy are just the conditions that ensure that C would follow whichever individual has the property F.

In the case of EE, the substitutional requirement demands that if an arbitrary name occurs in the existentially quantified line that triggers the assumption, some *new* arbitrary name must be chosen for the assumption. Intuitively, we want to show that C follows from a particular supposition, not from some stronger supposition. For example, if the line (∃y)Fay is eventually to be mentioned as an input to EE, the assumption that begins the proof of C must be Fab (never Faa). To show that C follows from the strong supposition that Faa would not be to show that it follows from the weaker assumption Fab. Again, intuitively, we do not want to permit the inference from (∃x)(∃y)Fxy to (∃x)Fxx.

In the case of UI, the substitutional requirement, intuitively enough, precludes the conclusion (x)(Fxa) from Faa, that is, it precludes the inference that everything is in a certain relation to **a** from the supposition that **a** is in the relation F to itself. We don't want a demonstration for some arbitrary individual that he loves himself, to become a pseudo-demonstration that everyone loves that individual, or that that individual loves everyone.

Exercise 5.1

For each of the following formulae, say whether it is a wff; if it is not a wff, say whether or not it is a propositional function. If it is a wff demonstrate that it is, by showing that it satisfies the definition.

(a) (x)Gxa

(b) (x)Gya

(c) (x)Gxy

(d) (∃x)(Fa ∧ Gx)

(e) (∃y)(Fa ∧ Gxy)

(f) (x)(∃y)(∃z)(Fy ∨ Gz → Hayz)

(g) (∃x)(∃y)(Fx ∨ Gy → (∃z)Hayz))

(h) (∃x)(∃y)(Fx → (z)(Gz → (∃x)Hxyz))

(i) (∃x)(Fx → (z)(Gz → (∃u)Hxyu))

(j) (∃y)((∃x)(Fx → (z)(Gz → (∃u)Hxyu))

Exercise 5.2

For each of the following supposed applications of UE, say whether it correct or incorrect. If it is incorrect, say why.

(a)	1	(1)	(x)(∃z)(Fxz ∧ Gxz)	A
	1	(2)	(∃z)(Faa ∧ Gaz)	1,UE
(b)	1	(1)	(x)(∃z)(Fxz ∧ Gxz)	A
	1	(2)	(∃z)(Faz ∧ Gbz)	1,UE
(c)	1	(1)	(x)(∃z)(Fxz ∧ Gxz)	A
	1	(2)	(∃z)(Fbz ∧ Gbz)	1,UE

Exercise 5.3

For each of the following supposed applications of EI, say whether it correct or incorrect. If it is incorrect, say why.

(a)	1	(1)	Fba	A
	1	(2)	(∃y)Fby	1,EI
(b)	1	(1)	Fba	A
	1	(2)	(∃x)Fxx	1,EI
(c)	1	(1)	Fba	A
	1	(2)	(∃y)Fya	1,EI
(d)	1	(1)	Fba	A
	1	(2)	(∃x)Fxb	1,EI
(e)	1	(1)	(∃x)Fxa	A
	1	(2)	(∃y)(∃x)Fxy	1,EI
(f)	1	(1)	(∃x)Fxa	A
	1	(2)	(∃y)(∃x)Fxx	1,EI

Exercise 5.4

For each of the following supposed applications of UI, say whether it correct or incorrect. If it is incorrect, say why. Assume that neither the arbitrary name a nor the arbitrary name b occurs in assumption (1)

<table>
<tr><td>(a)</td><td>1</td><td>(3)</td><td>Fab → (x)Gax</td><td></td></tr>
<tr><td></td><td>1</td><td>(4)</td><td>(y)(Fyb → (x)Gyx)</td><td>3,UI</td></tr>
<tr><td>(b)</td><td>1</td><td>(3)</td><td>Fab → (x)Gax</td><td></td></tr>
<tr><td></td><td>1</td><td>(4)</td><td>(x)(Fyb → (x)Gxx)</td><td>3,UI</td></tr>
<tr><td>(c)</td><td>1</td><td>(3)</td><td>Fab → (x)Gax</td><td></td></tr>
<tr><td></td><td>1</td><td>(4)</td><td>(y)(Fay → (x)Gax)</td><td>3,UI</td></tr>
<tr><td>(d)</td><td>1</td><td>(3)</td><td>Fab → (x)Gax</td><td></td></tr>
<tr><td></td><td>1</td><td>(4)</td><td>(y)(Fyy → (x)Gyx)</td><td>3,UI</td></tr>
</table>

Exercise 5.5

For each of the following pairs of wffs, say whether the second is an appropriate assumption for an application of EE in which the first of the pair is to be an input. If not say why.

(a)   (i)   (∃x)(Fxa ∧ (y)Gby)
        (ii)   Fba ∧ (y)Gby

(b)   (i)   (∃x)(Fxa ∧ (y)Gby)
        (ii)   Fca ∧ (y)Gcy

(c)   (i)   (∃x)(Fxa ∧ (y)Gby)
        (ii)   Fca ∧ (y)Gby

(d)   (i)   (∃x)(Fxa ∧ Gbx)
        (ii)   Fca ∧ Gbc

(e)   (i) $(\exists x)(Fxa \land Gbx)$
        (ii) $Fba \land Gbb$

(f)   (i) $(\exists x)(Fxa \land Gbx)$
        (ii) $Fbm \land Gbm$

5.3 L_q-theoremhood Preserving Operations

5.3.1 L_q-theorems

An L_q theorem is a wff that is L_q-provable from the empty set of assumptions. Since the set Φ_q of wffs includes the set Φ, and the R_q of rules of the system L_q includes the rules of L, all L-theorems are also L_q-theorems. Again, since the definition of *the conditional corresponding to a sequent* applies to all sequents, we may infer that every L_q sequent proved in the previous chapter has a corresponding conditional, and that that conditional is an L_q-theorem. Thus for example, corresponding to sequent 70, we have the theorem

91 $\vdash (x)(Fx \to Gx) \to ((x)Fx \to (x)Gx)$.

Corresponding to sequent **80**, we have the theorem

92 $\vdash (x)(Fx \to Gx) \leftrightarrow \neg(\exists x)(Fx \land \neg Gx)$.

Evidently, the definition of propositional uniform substitution can be extended in such a way as to give us yet another class of L_q-theorems. That is, the definition of Φ_q is such that for any wff A of Φ, the result of substituting in A an occurrence of any wff of Φ_q for every occurrence of any atom of A will be a wff of Φ_q. And, evidently as well, if A is an L-theorem, then the wff resulting from that uniform substitution will be an L_q-theorem, for we could just reconstruct the proof with that substitution made throughout. Thus, for example, from the proof of sequent **50** we would obtain a proof of

93 $\vdash Fa \lor \neg Fa$

by the simple expediency of replacing every occurrence of P in the former by an occurrence of Fa. Alternatively, we could have replaced every occurrence of P in that proof by an occurrence of Fm to obtain

94 ⊢ Fm ∨ ¬Fm

or (x)Fx to obtain

95 ⊢ (x)Fx ∨ ¬(x)Fx

or by an occurrence of any L_q wff to obtain an L_q theorem. Thus, uniform substitution of L_q wffs for L wffs in L_q-theorems inherited from L preserves theoremhood for L_q.

Two other classes of theorems result less directly from L-theorems. The proof of theorem 92 could be extended to a proof of

96 ⊢ (x)(Fx ∨ ¬Fx).

by an application of UI. Alternatively, the proof of theorem 92 or that of 93 could be extended to a proof of

97 ⊢ (∃x)(Fx ∨ ¬Fx)

by an application of EI. Thus UI and EI preserve L_q-theoremhood.

An application of uniform substitution in the propositional setting involves substitution for every occurrence of a propositional atom within a formula. Since in the language of quantificational logic, propositional atoms are understood as 0-place predicate letters, such an application amounts to a uniform replacement of all occurrences of a *0-place* predicate letter, that is, an *atomic propositional function in 0 variables*, within a wff by occurrences of *a propositional function in 0 variables*. The quantificational generalization of uniform substitution will involve the uniform replacement of all occurrences of an *atomic propositional function in n variables* by a *propositional function in n variables*. For example, in sequent **95**, we might replace every occurrence of Fx, an atomic propositional function in the single variable x with an occurrence of $(\exists y)Gxy$, a (non-atomic) propositional function in the single variable x. By doing so we would obtain

98   ⊢ $(x)((\exists y)Gxy \lor \neg(\exists y)Gxy)$.

Of course, this is just an example of such an operation applied to a quantificational theorem. The same wff could be shown to be a theorem by a uniform substitution in **50**

99   ⊢ $(\exists y)Gay \lor \neg(\exists y)Gay$

followed by an application of UI. The utility of a generalized quantificational account of uniform substitution will become apparent only when it is applied to distinctively quantificational sequents. Thus for example, we will expect that, in virtue of the theoremhood of **5.3.1**, we also have the theorem

100   ⊢ $(x)((\exists y)Hxy \to Gx) \to ((x)(\exists y)Hxy \to (x)Gx)$.

It is this wider range of applications of uniform substitution in the quantificational setting that demands a more general account of the operation.

5.3.2 Uniform substitution in L_q

In the following, it may be useful to think of a propositional function as representing a property together with an indication of its *arity* or *adicity*[2] Fx represents the one-place property **F**; Gxy the two-place property **G**, and so on. Every propositional function in n variables represents an n-place property.

Now recall from our account of the language of L_q, that the word *nominal* applies to anything which is either a term (an arbitrary name or a proper name) or an individual variable. So all of "a", "m", "x", and so

---

2. This is logicians' colloquial for the number of its places. A 2-place property is a binary property, that is, has arity 2; alternatively we can say that it is a dyadic property, that is has adicity 2. Typically logicians lapse into invention when they run out of Greek. So a 7-place property would be referred to as 7-ary rather than heptadic.

on, are called nominals. Our account of uniform substitution is the following:

Let A be a wff containing at least one occurrence of the predicate letter **P** followed by n nominals. Let $Q(v_1,\ldots,v_n)$ be a propositional function in n distinct variables such that no variable in a quantifier in $Q(v_1,\ldots,v_n)$ occurs in A. For any set of n nominals, $l_1,\ldots,l_n$, let $Q(l_1,\ldots,l_n)$ be the result of replacing each v_i by l_i in $Q(v_1,\ldots,v_n)$. Let A′ result from A by substituting $Q(l_1,\ldots,l_n)$ for each occurrence of $P(l_1,\ldots,l_n)$ in A. Then A′ results from A by *uniform substitution*.

A wff C is a *substitution instance* of a wff B when C is obtainable from B by a sequence of uniform substitutions.

Example:

Consider the theorem **5.3.1**

$$(x)(Fx \to Gx) \to ((x)Fx \to (x)Gx)$$

which for present purposes we shall call A. A contains two occurrences of the predicate letter F followed by the single nominal x.

$(\exists y)Ryz$ is a propositional function in the single individual variable z.

No variable occurring in a quantifier in that propositional function occurs in A.

$(\exists y)Ryx$ is the result of replacing the variable z in the propositional function by x.

Finally, the result of replacing every occurrence of Fx by $(\exists y)Ryx$ in A is the wff A′:

$$(x)((\exists y)Ryx \to Gx) \to ((x)(\exists y)Ryx \to (x)Gx)$$

A′ is obtained from A by substitution. A′ is a substitution-instance of A. We may obtain a second substitution-instance of A as follows:

A′ contains two occurrences of the predicate letter G followed by the single nominal x.

(z)(w)(∃y′)Hzwy′z′ is a propositional function in the single individual variable z′.

No variable occurring in a quantifier in that propositional function occurs in A′.

(z)(w)(∃y′)Hzwy′x is the result of replacing the variable z′ in the propositional function by x.

Finally, the result of replacing every occurrence of Gx by (z)(w)(∃y′)Hzwy′x in A′ is the wff A″:

$$(x)((\exists y)Ryx \to (z)(w)(\exists y')Hzwy'x)$$
$$\to ((x)(\exists y)Ryx \to (x)(z)(w)(\exists y')Hzwy'x)$$

A″ is obtained from A′ by substitution and A′ is obtained from A by substitution. Therefore A″ is a substitution-instance of A.

The operation of uniform substitution can of course be extended to ensembles of wffs.

Let Σ be an ensemble of wffs, at least one of which contains at least one occurrence of the predicate letter **P** followed by n nominals. Let $Q(v_1,\ldots,v_n)$ be a propositional function in n distinct variables such that no variable in a quantifier in $Q(v_1,\ldots,v_n)$ occurs in A. For any set of n nominals, $l_1,\ldots,l_n$, let $Q(l_1,\ldots,l_n)$ be the result of replacing each v_i by l_i in $Q(v_1,\ldots,v_n)$. Let Σ' result from Σ by substituting $Q(l_1,\ldots,l_n)$ for each occurrence of $P(l_1,\ldots,l_n)$ in the wffs of Σ. Then Σ' results from Σ by *uniform substitution*.

Now a sequent is an ensemble of wffs: the ensemble consisting of the wffs on the left of the ⊢ together with the wff on its right. We may therefore extend the notion of a substitution-instance from wffs to sequents. We can now state, though we will not here prove it, the expected metatheorem involving uniform substitution.

Metatheorem 5.0

If a sequent S has an L_q-proof, then every substitution-instance of S also has an L_q-proof.

Corollary 5.0.4

Every substitution-instance of an L_q-theorem is an L_q-theorem.

This metatheorem and its corollary afford us a technical advantage corresponding to the advantage conferred by their counterparts for propositional logic, namely that sequents previously proved can serve as derived, or non-primitive rules in proofs of later sequents. As in the propositional setting, we justify the application either by TI or TI(S), or by SI or SI(S), in the case of SI, mentioning the input wffs by line number, and in either case giving the number of the introduced theorem or sequent, or otherwise indicating where its proof can be found. Since theorems are proved from the empty ensemble, there will be no entry on the extreme left when a theorem is introduced. In the case of introduced sequents, the output wff is proved from whatever assumptions the input wffs are proved from. The quantificational uses of SI are illustrated in the following proofs.

101   (x)Fx ⊢ (x)(Gx → Fx)

1	(1)	(x)Fx	A
1	(2)	Fa	1,UE
1	(3)	Ga → Fa	2,SI(S) 56
1	(4)	(x)(Gx → Fx)	3,UI

102   (x)¬Fx ⊢ (x)(Fx → Gx)

1	(1)	(x)¬Fx	A
1	(2)	¬Fa	1,UE
1	(3)	Fa → Ga	2,SI(S) 57
1	(4)	(x)(Fx → Gx)	3,UI

103 $(\exists x)Fx \rightarrow (\exists x)Gx \vdash (\exists x)(Fx \rightarrow Gx)$

1	(1)	$(\exists x)Fx \rightarrow (\exists x)Gx$	A
2	(2)	$\neg(\exists x)(Fx \rightarrow Gx)$	A (RAA)
2	(3)	$(x)\neg(Fx \rightarrow Gx)$	2,SI(S) 4.9(d)
2	(4)	$\neg(Fa \rightarrow Ga)$	3,UE
2	(5)	$Fa \wedge \neg Ga$	4,SI(S) 3.5(g)
2	(6)	Fa	5,$\wedge$E
2	(7)	$\neg Ga$	5,$\wedge$E
2	(8)	$(\exists x)Fx$	6,EI
2	(9)	$(x)\neg Gx$	7,UI
1,2	(10)	$(\exists x)Gx$	1,8,MPP
1,2	(11)	$\neg(x)\neg Gx$	10,SI(S) 76
1,2	(12)	$(x)\neg Gx \wedge \neg(x)\neg Gx$	9,11,$\wedge$I
1	(13)	$\neg\neg(\exists x)(Fx \rightarrow Gx)$	2,12,RAA
1	(14)	$(\exists x)(Fx \rightarrow Gx)$	13,DN

5.3.3 Variable replacement and alphabetic variance

The individual variables of L_q can be thought of as mathematical counterparts of the definite pronouns *he, him, she, her, it* of English. There are, however, some striking differences. In the first place the individual variables are not distinguished either by gender (as *he* and *she* are) or by case (as *he* and *him* are). In the second place, there are infinitely many individual variables in the language of L_q. The multiplicity of them serves only to disambiguate grammatical references within single sentences. In L_q we don't have problems that would arise with such a sentence as *He was much taller than he was so that when he looked around him he could see nothing but a pair of boots which he recognized as his own.* Quantifiers correspond to the indefinite pronouns (someone, something, everyone, everything, anyone, anything) except that the quantifiers are not distinguished as between being personal or impersonal, and there are as many of them as there are individual variables. Anything that can be expressed with one finite set of quantifiers can be expressed with infinitely many other finite sets of quantifiers. Thus, for example, both $(x)(y)Fxy$ and $(x')(z)Fx'z$, 'say', for any particular interpretation of F, that everything F's everything. Such pairs of wffs are called *alphabetic variants*. Specifically, A and B are alphabetic variants if and only if B is obtainable from A by iterating the operation of *variable replacement.*

B results from A by variable replacement if and only if there are occurrences of the variable v in A, but no occurrences of the variable v', and B is the result of replacing every occurrence of v in A by an occurrence of v'.

(x)(y)Fxy and (x')(z)Fx'z are alphabetic variants because

  (1) (x)(y)Fxy contains no occurrences of the variable z, and (x)(z)Fxz is the result of replacing every occurrence of y by an occurrence of z,

and

  (2) (x)(z)Fxz contains no occurrences of the variable x', and (x')(z)Fx'z is the result of replacing every occurrence of x by an occurrence of x'.

The definition of variable replacement extends to sequents in the natural way: S' results from S by variable replacement if and only if there are occurrences of the variable v in at least one wff of S, but no occurrences of the variable v', and S' is the result of replacing every occurrence of v in any wff of S by an occurrence of v'. A sequent S' is an alphabetic variant of S if and only if S' can be obtained from S by a finite sequence of variable replacements.

Though we omit the proof, it is straightforward to demonstrate that the operation of variable replacement preserves theoremhood, and therefore provability. So if A is an L_q-theorem, and B is an alphabetic variant of A, then B too is an L_q-theorem, and if sequent S has a proof, and sequent S' is an alphabetic variant of S, then S' also has a proof. Accordingly, we extend the use of SI(S) and TI(S) to those cases in which the sequent or theorem introduced is an alphabetic variant of a sequent or theorem previously proved. So, for example, SI(S) 103 can be given as the justification for the extension of a proof from one having as a line the wff $(\exists y)Fy \rightarrow (\exists y)Gy$ to a proof having as a line the wff $(\exists y)(Fy \rightarrow Gy)$.

5.3.4 Term replacement

If wffs A and B are alphabetic variants, then $\vdash A \leftrightarrow B$. That variable replacement preserves theoremhood is a consequence of that fact. That is, a proof of A as a theorem can be extended by TI (or TI(S)) and an application of MPP (or $\wedge$E and MPP) to a proof of B as a theorem. In the case of sequents, if S and S$'$ are alphabetic variants and S has a proof, then from the premisses of S$'$, the premisses of S can be proved, then that proof extended to a proof of the conclusion of S, and that proof extended to a proof of the conclusion of S$'$. This is so because there is a one-to-one correspondence between the premisses and conclusion of S and the premisses and conclusion of S$'$, the pairs of corresponding wffs being interderivable. But we must distinguish the case in which $\vdash A \leftrightarrow B$ from the weaker case of wffs A and B for which there is no such single theorem, but for which $\vdash A$ if and only if $\vdash B$. Consider sequent **62**

$$(x)(Fx \rightarrow Gx), Fm \vdash Gm.$$

Replacing every occurrence of m by an occurrence of n, we obtain the sequent

104 $(x)(Fx \rightarrow Gx), Fn \vdash Gn.$

Evidently neither Fn $\leftrightarrow$ Fm nor Gn $\leftrightarrow$ Gm is a theorem of L_q, and yet the proof of sequent **103** will exactly parallel the proof of sequent **62**. Indeed we can infer that sequent **103** has a proof from the fact that sequent **62** has a proof. And again we could infer that the wff of

105 $\vdash (x)(Fx \rightarrow Gx) \rightarrow (Fa \rightarrow Ga)$

is a theorem from the fact that

106 $\vdash (x)(Fx \rightarrow Gx) \rightarrow (Fm \rightarrow Gm).$

The operation by which we obtain sequent **103** from sequent **62** or Theorem **104** from Theorem **105** is called *term replacement*: A$'$ is obtained from A by term replacement if and only if A is a wff with at least one occurrence of the term t and no occurrences of the term t', and A' is the result of replacing every occurrence of t in A by t'. Sequent S' is obtained from sequent S by term replacement if and only if the wffs of S' are in one-to-one correspondence to those of S, and each wff is either identical to its correspondent or obtained from it by term-replacement. Notice that term replacement encompasses indifferently both replacements of proper names with proper or arbitrary ones and replacements of arbitrary names with arbitrary or proper ones. Term replacement preserves provability of sequents and theoremhood of wffs.

5.4 Models for Quantificational Logic

In logic a *model* is any device by which a truth-value is obtained for every wff.

5.4.1 Propositional models revisited

Recall that the atomic sentences of propositional logic, (*i.e.* the propositional variables) have no internal structure. Therefore, for atomic *propositional* wffs, internal structure can play no role in the determination of their truth values.

All that is required, in fact all that is possible, for a propositional model is an assignment of truth values to all of the propositional variables.

The truth values of all non-atomic wffs can then be computed by reference to the matrices for $\neg$, $\rightarrow$, $\vee$, and $\wedge$.

When we construct a truth table for a wff of n propositional variables, we are setting out 2^n distinguishable classes of models for a propositional language.

Each row of the table may be taken to represent all assignments which differ only with respect to the values that they assign to propositional variables having no occurrences in the present wff.

For example in the first row of the table for the wff:

$$P \lor \neg P \to (Q \to R) \lor \neg(Q \to R)$$

We find represented the class of assignments that give the value **1** to P, Q, and R, though they may differ in the values that they give to other propositional variables P′, Q′, R′ and so on.

Atoms

Quantificational atoms have internal structure. That is, they consist of a predicate symbol followed by some finite number of terms. Accordingly, quantificational models must take this internal structure into account, by *interpreting* both predicate symbols and terms. The fundamental device by which a quantificational model does this is by reference to a non-empty set of objects called a *domain of individuals*.

One-place predicate symbols

Terms are interpreted as *names of individuals* through being assigned to members of the domain. Each one-place predicate symbol P is interpreted as representing some *property* of individuals. In effect, the model assigns the predicate symbol to some set of individuals in the domain, which intuitively we can think of as the set of individuals that have the property it is taken to represent.

Many-place predicate symbols

In general, n-place predicates are interpreted as *n-ary relations*, which are identified with a set of n-tuples of individuals (thought of as the set of n-tuples of individuals which are in the relation.)

Example:

Consider the sentence, Fa ∧ Gab.

One model might take the set of natural numbers as its domain. It might interpret F as the property of being even, and G as the relation of being greater. Now suppose that a is interpreted as the number 5, and b is interpreted as the number 4.

Then the sentence Fa will receive the value 0, since 5 is not even, and consequently the whole sentence Fa ∧ Gab will receive the value 0 by the truth table for ∧.

A different model on the same domain, which interpreted a as the number 4 and b as the number 3 would give the same sentence the truth value 1, since 4 is even and 4 is greater than 3. A wff is said to be *satisfied* in a model which gives it the value 1.

5.4.2 Truth values of quantified sentences

A model will assign the wff

$$(x)(....x....)$$

the value 1 if *every* individual in the domain has the property expressed by the propositional function (...x...). A model will assign the wff

$$(\exists x)(....x....)$$

the value 1 if *some* individual in the domain has the property expressed by the propositional function (...x...).

5.4.3 Validity

A wff is *valid on domain D* if and only if it receives the value 1 in every quantificational model on the domain D. It is said to be *universally valid* if and only if it is valid on every domain. If a quantificational wff has only zero- or one-place predicate symbols, then it is universally valid if and only if it receives the value 1 in all models on a domain of 2^n individuals if it contains n distinct one-place predicate symbols. (The set of wffs containing only occurrences of one-place predicate symbols is called *the monadic fragment* of quantificational language.) The reason is analogous to the reason why a propositional wff containing n propositional variables is valid if it takes the value 1 in each of the 2^n rows of its truth table. Just as n propositional variables permit us to distinguish no more than 2^n classes of propositional models, so n one-place properties will enable us to distinguish no more than 2^n individuals. Even if there were more individuals in the domain, we would not be able to distinguish them from the first 2^n.

Since the set of all finite models is enumerable, we can determine for any wff of the monadic fragment whether or not it is valid. We simply look at all possible interpretations for a domain of one individual, then at all possible interpretations for a domain of two individuals and so on. If a formula is invalid then it must fail in some interpretation on a domain having no more than 2^n individuals. If it receives the value one in all models with no more than 2^n individuals then it is valid.

The case is much different for wffs outside the monadic fragment. Consider the following triad of wffs:

$(x)\neg Rxx$ (R is irreflexive)

$(x)(y)(z)(Rxy \wedge Ryz \rightarrow Rxz)$ (R is transitive)

$(x)(\exists y)Rxy$ (R is serial)

Exercise 5.6

Can the conjunction of that triad of wffs be satisfied in a finite model?

5.5 Determining validity

Consider a model whose domain has two individuals, m and n. The wff:

$$(x)(y)(Ax \rightarrow (By \vee Cy))$$

will be true in a model if and only if the property expressed by the propositional function:

$$(y)(Ax \rightarrow (By \vee Cy))$$

in that model is true of every individual in the domain of the model, that is, if and only if:

$$(y)(Am \rightarrow (By \vee Cy)) \wedge (y)(An \rightarrow (By \vee Cy))$$

But that property will be true of every individual if and only if the property expressed by the propositional function:

$$Am \rightarrow (By \vee Cy)$$

in that model is true of every individual in the domain of the model and likewise the property expressed by the propositional function:

$$An \rightarrow (By \vee Cy)$$

in that model is true of every individual in the domain of the model, that is if and only if:

$$((Am \rightarrow (Bm \vee Cm)) \wedge (Am \rightarrow (Bn \vee Cn))) \wedge$$
$$((An \rightarrow (Bm \vee Cm)) \wedge (An \rightarrow (Bn \vee Cn)))$$

is true in that model. The wff

$$(z)(((y)By \vee (y)Cy) \rightarrow Dz)$$

will be true if and only if the propery expressed by the propositional function:

$$(y)By \vee (y)Cy \rightarrow Dz$$

in that model is true of every individual, that is if and only if:

$$((y)By \vee (y)Cy \rightarrow Dm) \wedge ((y)By \vee (y)Cy \rightarrow Dn),$$

which in turn will be true if and only if:

$$((Bm \wedge Bn) \vee (Cm \wedge Cn) \rightarrow Dm) \wedge ((Bm \wedge Bn) \vee (Cm \wedge Cn) \rightarrow Dn)$$

is true in the model. The wff:

$$(\exists x)(\exists z)(Ax \rightarrow Dz)$$

will be true if and only if the property expressed by the propositional function:

$$(\exists z)(Ax \to Dz)$$

is true of some individual in the domain of the model, that is if and only if:

$$(\exists z)(Am \to Dz) \vee (\exists z)(Am \to Dz)$$

which will be true if and only if:

$$((Am \to Dm) \vee (Am \to Dn)) \vee ((An \to Dm) \vee (An \to Dn))$$

Thus the question whether the argument:

$$(x)(y)(Ax \to (By \vee Cy));$$
$$(z)(((y)By \vee (y)Cy) \to Dz);$$
$$\text{therefore, } (\exists x)(\exists x)(Ax \to Dz)$$

can be shown to be invalid by a model on a domain of two individuals can be expressed as the question: is there an assignment of values which will make the first two of the following sentences true, and the third of them false?

$((Am \to (Bm \vee Cm)) \wedge (Am \to (Bn \vee Cn))) \wedge ((An \to (Bm \vee Cm)) \wedge (An \to (Bn \vee Cn)))$

$((Bm \wedge Bn) \vee (Cm \wedge Cn) \to Dm) \wedge ((Bm \wedge Bn) \vee (Cm \wedge Cn) \to Dn)$

$((Am \to Dm) \vee (Am \to Dn)) \vee ((An \to Dm) \vee (An \to Dn))$

The question whether that quantificational argument can be shown invalid by a model on a singleton domain (or "unit domain" = "domain of one individual") would be expressible as the corresponding question about the triad of sentences:

$$Am \to Bm \vee Cm$$
$$Bm \vee Cm \to Dm$$
$$Am \to Dm.$$

Evidently, no interpretation of A, B and C in a model on a singleton domain will show the original argument to be invalid, since any substitution instance of the sequent

$$P \to Q, Q \to R \vdash P \to R$$

is valid, including the substitution instance in question.

But the original quantificational argument can be shown to be invalid by a model on a domain of two individuals. Simply interpret A, B, and C in such a way that:

$$Am = An = Bm = Cn = 1$$

and

$$Bn = Cm = Dm = Dn = 0.$$

5.5.1 A second example

Consider the sequent:

$$(x)(\exists y)(Ex \to Fy), (\exists y)(z)(Fy \to \neg Gz) \vdash (x)(z)(\neg Ex \to Gz)$$

on a domain of two individuals m and n, the relevant sequent would be:

$$((Em \to Fm) \lor (Em \to Fn)) \land ((En \to Fm) \lor (En \to Fn)),$$
$$((Fm \to \neg Gm) \land (Fm \to \neg Gn)) \lor ((Fn \to \neg Gm) \land (Fn \to \neg Gn)) \vdash$$
$$((\neg Em \to Gm) \land (\neg Em \to Gn)) \land ((\neg En \to Gm) \land (\neg En \to Gn)).$$

This sequent is shown to be invalid by an assignment of the value 0 to every atom.

The consideration of a model on a two-individual domain is useful as an indication of how the translation goes, but to prove the sequent invalid, we need not have been so ambitious. For on a domain of one individual m, the relevant sequent will be:

$$Em \to Fm, Fm \to \neg Gm \vdash \neg Em \to Gm$$

which is obviously also invalid.

Exercise 5.7

Each of the following claims an entailment. For each claim, use the **Simon** Quantificational Model Search (QMS) Editor to construct propositional expansions for small domains beginning with domain size 1. Test each using the **Simon** QMS Editor and say whether or not the claim is correct.

(a) (x)Fx ⊨ (∃x)Fx
(b) (∃x)Fx ⊨ (x)Fx
(c) (x)(∃y)(Fx → Fy) ⊨ (∃y)(x)(Fx → Fy)
(d) (∃y)(x)(Fx → Fy) ⊨ (x)(∃y)(Fx → Fy)
(e) (x)(∃y)(Fx → Gy) ⊨ (∃y)(x)(Fx → Gy)
(f) (∃y)(x)(Fx → Gy) ⊨ (x)(∃y)(Fx → Gy)
(g) (x)(Fx → Gx) ⊨ (∃x)(Fx ∧ Gx)
(h) (∃x)(Fx ∧ Gx) ⊨ (x)(Fx → Gx)
(i) (x)(Fx → Gx) ⊨ (∃x)(Fx ∨ Gx)
(j) (∃x)(Fx ∨ Gx) ⊨ (x)(Fx → Gx)
(k) (x)Fx → (x)Gx ⊨ (x)(Fx → Gx)
(l) (x)(Fx → Gx) ⊨ (x)Fx → (x)Gx
(m) (x)Fx ∨ (x)Gx ⊨ (x)(Fx ∨ Gx)
(n) (x)(Fx ∨ Gx) ⊨ (x)Fx ∨ (x)Gx
(o) (x)(Fx → Gx) ⊨ ¬(∃x)Gx → ¬(∃x)Fx
(p) (x)(Fx → Gx) ⊨ ¬(∃x)Gx → (∃x)¬Fx
(q) (x)Fx → (x)Gx ⊨ (∃x)(Fx → Gx)
(r) (x)Fx → (x)Gx ⊨ (∃x)Fx → (∃x)Gx
(s) (x)(Fx → Gx) ⊨ (∃x)Fx → (∃x)Gx
(t) (∃x)Fx → (∃x)Gx ⊨ (x)(Fx → Gx)
(u) (∃x)Fx → (∃x)Gx ⊨ (x)Fx → (x)Gx
(v) (x)(∃y)(Fx ∧ Gy) ⊨ (∃y)(x)(Fx ∧ Gy)
(w) (∃y)(x)(Fx ∧ Gy) ⊨ (x)(∃y)(Fx ∧ Gy)

5.6 First-order Theories

The subject matter of this and the previous chapter is sometimes referred to as *first-order logic*. The label is meant to distinguish a system of logic which permits quantification only over individuals from one which also permits quantification over properties. Systems which permit quantification over properties of individuals are called systems of *second-order logic*; those which permit quantification over properties of properties of individuals are called systems of *third-order logic*. Collectively, such systems are referred to as systems of *higher-order logic*. Systems such as the one presented here are also referred to as *The Lower Predicate Calculus* or *LPC*. Natural language use traverses these orders without difficulty, as when we say that a person inherits some of the traits of his mother and some of the traits of his father. To represent such a sentence we might even have recourse to second-order properties such as that of *being a trait* or even to mixed-order relational properties such as *being a trait of*:

$(x)(y)(z)(Px \rightarrow (Py \rightarrow (Pz \rightarrow (Mxz \rightarrow (Fyz \rightarrow (\exists\varphi)(T\varphi \wedge (\varphi x \wedge \varphi z)) \wedge (\exists\varphi)(T\varphi \wedge (\varphi y \wedge \varphi z)))))))$

in which φ is a predicate variable, T is the second-order property of being a trait, and P, M, and F are the first-order properties of being a person, being a mother, and being a father, respectively. While we generally encounter no problems in natural-language uses of higher-order predicates, problems attend the formulations of systems of higher-order systems that represent them. Consider only that the property of *being a property* is a property not only of properties of individuals, but also of properties of properties of individuals and so on, and you have some notion of the richness of the conceptual terrain. Consider that *not being a property* is a property of non-properties, and you begin to suspect the perils that it holds. Consider further that either the property *does not have itself* or the property *has itself* can be a property of a property. For example, the property *not being a property* does not have itself, since it is a property, and the property *being a property* has itself, since being a property is a property of every property. We need not venture much further to be face to face with a curious difficulty: does the property *does not have itself* have itself or not? Evidently, if it does, then it does not; if it does not, then it does.

While the set of rules which we have adopted is in fact complete for the intended interpretation of quantificational language (which is to say that all valid quantificational wffs are theorems, and all valid quantificational sequents have proofs), the study of the quantificational rules that we adopted represents the barest beginning of a study of applied first-order logic.

First-order logic provides the foundation of special theories called *first-order theories* (fo theories).

A fo theory enriches the formal quantificational system in two ways:

(a) It adds to the stock of symbols of the language some set of special *predicate symbols*

(b) It adds to the stock of primitive sequents (rules) some additional sequents intended to capture the inferential properties of the special predicate symbols for their intended interpretation.

Example:

For a first-order theory of the relation of *strict relative greatness* on the natural numbers, we would expect to add to the stock of predicate symbols the two-place predicate symbol '<'.

We would expect new primitive sequents which were sufficient to permit the proof of such theorems as:

$\vdash$ (x)¬<xx

$\vdash$ (x)(y)(z)(<xy ∧ <yz → <xz)

$\vdash$ (x)(∃y)<xy.

In fact these might be added as primitive sequents.

5.6.1 The first-order theory of identity

It is worth remarking that the earliest theory of identity that has been proposed was a *second-order* theory proposed by Gottfried Wilhelm Leibniz (1646-1716) in a pair of principles called *The principle of the identity of indiscernibles* and *The law of identity*. The former principle stated that:

Individuals having all of their properties in common are identical.

Stated in current quantificational notation, this is the principle:

$$(x)(y)((\delta)(\delta x \leftrightarrow \delta y) \rightarrow x = y)$$

It is second-order because it has occurrences of the variable, δ, which ranges over properties of individuals, rather than over individuals. The quantifier '(δ)' is read, 'for every property δ...'.

In second-order logic we could simply *define* identity by this means:

$$t = t' =_{df} (\delta)(\delta x \leftrightarrow \delta y)$$

The second of Leibniz's principles, the law of identity, asserted that every individual was identical to itself.

$$(x)x = x$$

In first-order logic, we add the 2-place predicate symbol '=' to our stock of predicate symbols by adding a clause to our definition of Φ, that also makes '=' explicitly binary.

If t and t' are terms then $t = t'$ is a wff.

At the same time we make it explicit that '=' will be written in its usual *infix* position. (So we do not write '=tt'' but rather 't = t'.)

We then add to our stock of rules, rules for the introduction and elimination of identity.

Identity Introduction (=**I**)

For any term t, $t = t$ may be introduced as a line of a proof. It requires no assumptions.

Identity Elimination (=**E**)

Let 's' and 't' be terms, and A(t) be a wff containing occurrences of **t**. Let A(s) be the result of replacing at least one occurrence of 't' in A(t) by 's'. Then from A(t) and t = s, we may infer A(s). The conclusion requires the assumptions required by the premisses.

5.6.2 Sequents involving identity

Notice that in the statement of =E, the order of "t" and "s" in the identity premiss is important, the rule does not permit us to write the line A(s) given only the lines A(t) and s = t. However, we can prove a sequent which will enable us to commute identities:

107 $a = b \vdash b = a$

1	(1)	$a = b$	A
	(2)	$a = a$	=I
1	(3)	$b = a$	1,2,=E

108 $a = b \wedge b = c \vdash a = c$

1	(1)	$a = b \wedge b = c$	A
1	(2)	$a = b$	1,∧E
1	(3)	$b = c$	1,∧E
1	(4)	$a = c$	2,3,=E

109 $Fa \dashv\vdash (\exists x)(x = a \wedge Fx)$

(⊢)

1	(1)	Fa	A
	(2)	$a = a$	=I
1	(3)	$a = a \wedge Fa$	1,2,∧I
1	(4)	$(\exists x)(x = a \wedge Fx)$	3,EI

(⊣)

1	(1)	$(\exists x)(x = a \wedge Fx)$	A
2	(2)	$b = a \wedge Fb$	A (EE)
2	(3)	$b = a$	2,∧E
2	(4)	Fb	2,∧E
2	(5)	Fa	3,4,=E
1	(6)	Fa	1,2,5,EE

110 ⊢ (x)x = x

	(1)	a = a	=I
	(2)	(x)x = x	1,UI

111 ⊢ (x)(y)(x = y → y = x)

1	(1)	a = b	A (CP)
	(2)	a = a	=I
1	(3)	b = a	1,2,=E
	(4)	a = b → b = a	1,3,CP
	(5)	(y)(a = y → y = a)	4,UI
	(6)	(x)(y)(x = y → y = x)	5,UI

112 ⊢ (x)(y)(z)(x = y ∧ y = z → x = z)

1	(1)	a = b ∧ b = c	A(CP)
1	(2)	a = b	1,∧E
1	(3)	b = c	1,∧E
1	(4)	a = c	2,3,=E
	(5)	a = b ∧ b = c → a = c	1,4,CP
	(6)	(z)(a = b ∧ b = z → a = z)	5,UI
	(7)	(y)(z)(a = y ∧ y = z → a = z)	6,UI
	(8)	(x)(y)(z)(x = y ∧ y = z → x = z)	7,UI

5.6.3 Expressing distinctness and number

The wff

$$(\exists x)Fx \land (\exists y)Fy$$

does not give us the resources to express the existence of two individuals having the property F, since, as we know from sequent **79**, that wff is provable from the wff

$$(\exists x)(Fx).$$

How then do we say that there are *at least* two individuals which have the property F? The identity predicate together with negation gives us the means of expressing distinctness. Thus, to say in the language of L_q with identity that two individuals have the property F, we add to the quoted conjunction a conjunct expressing the non-identity of x and y:

$$(\exists x)(\exists y)((Fx \wedge Fy) \wedge \neg x = y)$$

Again, to say that there are at least three such individuals, we must assert the identity of three individuals no pair of which are identical.

$$(\exists x)(\exists y)(\exists z)(((Fx \wedge Fy) \wedge Fz) \wedge ((\neg x = y \wedge \neg(x = z)) \wedge \neg(y = z)))$$

and so on. Howsoever cumbrously, we can express any finite lower bound on the number of individuals having any given first-order property.

With recourse to identity, we can also express any finite upper bound on the number of individuals having a first-order property. To say that there is *at most* one individual having the property F, we say in effect that any pair of pronouns occurring in true attributions of the property must refer to the same individual.

$$(x)(y)(Fx \wedge Fy \rightarrow x = y)$$

To say that there are at most two such individuals, we say, again 'in effect' that for any triple of such pronouns, two of them must refer to the same individual.

$$(x)(y)(z)((Fx \wedge Fy) \wedge Fz \rightarrow (x = y \vee x = z) \vee y = z)$$

Finally, since we have the means to express both any finite lower bound, and any finite upper bound on the number of individuals having F, we also have the means of fixing both at any common n, that is, of saying, for any n, that there are exactly n individuals having the property F. To say that there is *exactly* one individual having the property F we say that there is at least one such individual, and that there is at most one:

$$(\exists x)(Fx \wedge (y)(Fy \rightarrow x = y)).$$

To say that there are exactly two such individuals we say that there are at least two and that there are at most two

$$(\exists x)(\exists y)(((Fx \wedge Fy) \wedge \neg x = y) \wedge (z)(Fz \rightarrow x = z \vee y = z))$$

and so on.

5.6.4 Definite Descriptions

The expression *definite description* is due to Bertrand Russell, who distinguished between such *denoting phrases* as *an author of poems*, which he labelled *indefinite descriptions*, and noun phrases of the form *the X* such as *the composer of 'Strawberry Fields Forever'* or *the lady on the corner*, which he labelled *definite descriptions*. In supposing that L_q with identity provides any adequate analysis of such natural language expressions, we are straying into one of those delusionary syndromes where only philosophers feel entirely at ease. But it does no harm to observe the conventional first-order representation of definite descriptions for purely callisthenic purposes. Definite descriptions are conventionally taken to imply both existence and uniqueness. Thus to say that *the individual having the property F* also has the property **G**, we say, in the manner described earlier, that there exists exactly one individual having F, and to that conjoin the ascription of G to that individual, as:

$$(\exists x)((Fx \wedge (y)(Fy \rightarrow x = y) \wedge Gx).$$

To say that the English sentence *The woman on the corner has slipped a disk* implies that there is exactly one lady on the corner is perhaps to apply the strong language of implication where some weaker verb such as *suggests* might better capture the vagueness of our intuitions. We might feel that it is too strong to say that the sentence is false in virtue of the presence of a police woman who has rushed to her assistance. And in any case we might continue to refer to someone as the woman on the corner having so identified her when she was alone. Again, consider the sentences (all from *The Globe and Mail*)

> Constable Strongquill's friends and co-workers spent the weekend coming to grips with the death of the likable Mountie. . . (2001-12-24),

> The officer had five other children all from other marriages. (2001-12-27)

> The twenty-seven year old woman and her infant daughter, Korrie, were with family and friends through Christmas. (Ibid.)

which we do not take to imply that after that constable's untimely death, there remain no agreeable RCMP officers or that there are no other officers, or that there is only one twenty-seven year old woman. And again, the *the dog* of *The dog is man's best friend* may suggest that exactly one species supplies that market, but that is not the role of the *the*. And again we will hardly take the sentence *The perfect mousetrap does not exist* to imply the *existence* of a representative mousetrap, even if we take it to imply that there could be only one such species. Finally, we do not take the sentence *The woman on the corner does not exist* to be true in virtue of there being no unique such woman. As in the case of the connectives, we recognize that the sentence is representable in such and such a way in our first-order language by our understanding it in a certain way, not by our noticing a certain configuration of words. It should be borne in mind that the prescribed representation of definite descriptions was originally introduced for mathematical representations, the smallest prime number, the hypotenuse of a given right triangle, and so on, not as an analysis of the word *the*. In the case of many non-mathematical definite descriptions, one important clue to their appropriate representation will be their occurring in an exercise of a logic text.

5.6.5 Superlatives

Superlatives, like definite descriptions, are treated with an almost unseemly briskness in their conventional logical treatment, usually as though in the light of a brief discussion about definite descriptions, they can be seen as particular instances of the same linguistic construction. Certainly superlatives (*the richest woman in Europe, the tallest building in this country, the best instructor* and so on) are often formed with the definite article. But if they were treated in the manner of definite descriptions, we would require a clause that makes uniquenes explicit. For 'Yue is the tallest student', we would have to say, for example, 'Yue is taller than any other student, *and no other student is.*' Nevertheless, we do not require the addition such a clause. Instead, to represent the F-est we use a (typically) two-place predicate symbol F taken to represent the comparative form, F-er than and say that the object is F-er than any other object in the class of comparison. So, for example, to represent the

English sentence *The best instructor is a student* we might construct the following:

$$(\exists x)((Ix \wedge Sx) \wedge (y)(Iy \rightarrow (\neg x = y \rightarrow Bxy))).$$

We do not feel it necessary to add a clause saying that every other instructor is not better than every other instructor, as in

$$(\exists x)((Ix \wedge Sx) \wedge (y)(Iy \rightarrow (\neg x = y \rightarrow Bxy \wedge \neg(z)(Iz \wedge \neg z = y \rightarrow Byz)))\ ).$$

Why? The reason is presumably that we accept that if m is F-er than n, then n is not F-er than m. But that principle amounts to an additional assumption, in the case of B, the assumption:

$$(x)(y)(Bxy \rightarrow \neg Byx).$$

Now that assumption might be true of all comparatives, and if it is, then it certainly guarantees uniqueness. But then it ought either to proved from some stipulated principles of comparatives, or itself stipulated explicitly as an unproven postulate. In this matter as in others, we acquiese in standard practice. In the following we take the liberty of making the assumption that comparatives have this property, (called *asymmetry*), and we therefore represent superlatives in the fashion of the earlier illustration.

Exercise 5.8

Represent each of the following arguments as a quantificational sequent using the notation provided, and use **Simon**'s proof editor to construct a proof for it.

(a) The God who begat Polyphemus begets only cyclopes. Therefore Polyphemus is a cyclops. (Bxy: x begets y; Gx: x is a god; Cx: x is a cyclops; m: Polyphemus)

(b) The man who broke the bank at Monte Carlo is very rich. Therefore, every man who broke the bank at Monte Carlo is very rich. (Mx: x is a man; Bx: x broke the bank at Monte Carlo; Rx: x is very rich)

(c) The yappiest maltipoo in the kennel is incontinent. No incontinent maltipoo is a welcome pet. Therefore the yappiest maltipoo in the kennel is not a welcome pet. (Mx: x is a

maltipoo; Ix: x is incontinent; Kx: x is in the kennel; Wx: x is
a welcome pet; Yxy: x is yappier than y)

(d)   The fastest human knitter is a Scot. One person can knit faster
      than another only if more nimble. Scots eat haggis with
      pleasure. Therefore anyone to whom the eating of haggis
      affords no pleasure is less nimble than someone or other. (**P**x:
      x is a person; **K**xy: x can knit faster than y; **S**x: x is a Scot;
      Nxy: x is nimbler than y; **H**x: x gets pleasure from eating
      haggis.)

(e)   No crook pays higher bribes than those paid by a richer crook.
      Only crooks pay bribes. No two crooks are equally rich. No
      two crooks give the same bribe. It isn't a bribe unless it's
      paid. Therefore, if there is a highest bribe, then there is a
      richest crook. (Cx: x is a crook; Bx: x is a bribe; Pxy: x pays
      y; Rxy: x is richer than y; Hxy: x is higher than y)

(f)   Richer crooks pay higher bribes than poorer crooks. Only
      crooks pay bribes. All bribes are paid by crooks. Therefore, if
      there is a highest bribe, then there is a richest crook. (Cx: x is
      a crook; Bx: x is a bribe; Pxy: x pays y; Rxy: x is richer than
      y; Hxy: x is higher than y)

(g)   Richer crooks pay higher bribes than poorer crooks. Only
      crooks pay bribes. All crooks resort to bribery. No two crooks
      are equally rich. No two crooks give the same bribe. Of any
      two bribes, at most one is higher than the other. Therefore, if
      a highest bribe is paid, then there is a richest crook. (Cx: x is
      a crook; Bx: x is a bribe; Pxy: x pays y; Rxy: x is richer than
      y; Hxy: x is higher than y)

(h)   Either Nelson's death was unavenged or some French ship
      was capsized. For every French ship capsized, there was a
      happy English tar. If one English tar was happy so were they
      all. Therefore if any English tar was unhappy, then Nelson's
      death was unavenged. (n: Nelson; Ax: x's death is avenged;
      Fx: x is French; Sx: x is a ship; Cx: x is capsized; Tx: x is a
      tar; Ex: x is English; Hx: x is happy)

(i)   Every song was a hit. There was at most one hit. There was at
      least one song. Therefore there was exactly one song. (Sx: x
      was a song; Hx: x was a hit)

(j)   Every hit was a song. There was at least one hit. There was at
      most one song. Therefore there was exactly one song. (Sx: x
      was a song; Hx: x was a hit)

(k)   Marcia and Naomi were the only women at the ball who had
      recently bathed. Every woman at the ball who survived had
      recently bathed. Marcia, unfortunately, did not survive. If any
      woman at the ball had recently bathed, then some woman at

the ball who had bathed must have survived. All who had bathed fainted. Therefore the woman at the ball who survived fainted. (m: Marcia; n: Naomi; Wx: x was a woman at the ball; Bx: x had recently bathed; Sx: x survived; Fx: x fainted etc.)

(l) At most two acupuncturists are barristers. All county-court judges are barristers. At least two county-court judges are acupuncturists. Therefore exactly two barristers are acupuncturists. (Ax: x is an acupuncturist; Bx: x is a barrister; Cx: x is a county-court judge.)

(m) Anyone who has swum in English Bay is hardier than anyone who has not. Of all my acquaintances, the stupidest alone has swum in English Bay. All my acquaintances are ill. Therefore the hardiest of my acquaintances is ill. (Ax: x is an acquaintance of mine; Ex: x has swum in English Bay; Ix: x is ill; Hxy: x is hardier than y; Sxy: x is stupider than y.)

(n) For anyone less happy than someone, there is someone than whom he is less happy who is himself less happy than no one. Therefore, if for every person who isn't rich there is a happier person who is, and for every untalented person there is a happier person with talent, then for every person who is neither rich nor talented, there is a happier person who is both. (Hxy: x is happier than y; Rx: x is rich; Tx: x is talented.)

(o) For any three objects, if the first is between the second and the third then either the second is between the first and the first, or the third is between the first and the first. Therefore, for any three objects, if the first is between the second and the third then either the first is between the second and the second or the first is between the third and the third. (Bxyz: x is between y and z.)[3]

---

3. Try to devise a space in which betweenness non-trivially satisfies both the premiss and the conclusion. As a clue consider points on a circle, (or ball-bearings in a hula-hoop) and understand **a** to be between **b** and **c** iff (from a fixed point of view) there is a clockwise path from **a** to **b** that does not pass through (or is not blocked by) **c**, and a counterclockwise path from **a** to **c** that does not pass through (is not blocked by) **b**. Evidently in such a space not only will we have both **Babb** and **Bacc** if we have **Babc**; we will also have **Baaa**. A more interesting space would be one which did not in general permit the *or* of the conclusion to be replaced by an *and* and would not in general give us (x)Bxxx.

5.7 First-order Theories of Relations

A fo theory of a single binary relation **R** will add to the definition of wffs, the additional clause:

If t and t′ are terms then **R**tt′ is a well-formed formula.

It will add to the basic set of rules, now taken to include

> (a) a complete set of propositional rules,
> (b) the quantificational rules, and
> (c) the rules of identity,

some additional principles (rules and theorems) which express the inferentially essential features of the relation whose theory is being propounded.

What additional principles are introduced will be determined by reference to the intended models for the theory. Recall that a model interprets the two-place predicate R as some or other binary relation on the domain of the model. For example, if one intended model of the first-order theory of R had the set of natural numbers as domain, and the 'strictly less than' relation ($<$) as the intended interpretation of R, then we would expect to find among the theorems (primitive or derived) of R, the following wffs:

> $(x)\neg Rxx$
> $(x)(\exists y)Rxy$
> $(x)(y)(Rxy \rightarrow \neg Ryx)$
> $(x)(y)(z)(Rxy \rightarrow (Ryz \rightarrow Rxz))$
> $(x)(y)(\neg x = y \rightarrow Rxy \vee Ryx)$

It is usual and convenient to think of these wffs as expressing *properties of* R. Because the properties are expressible in first-order wffs, they are called *first-order properties of R*. However, not all properties of a binary relation are first-order properties. The property expressed by the second-order wff:

$$(x)(\theta)((y)(Rxy \rightarrow (\exists z)(Ryz \wedge \theta z)) \rightarrow (\exists y)(Rxy \wedge (z)(Ryz \rightarrow \theta z)))$$

cannot be expressed in any wff of a first-order language. But many of the fo properties of relations are expressed by first-order wffs which recur frequently enough to be familiar. These have been given suitable mnemonic names:

Examples:

(x)¬Rxx (*irreflexivity*: R is *irreflexive.*)
(x)(∃y)Rxy (*seriality*: R is *serial.*)
(x)(y)(Rxy → ¬Ryx) (*asymmetry*: R is *asymmetric.*}
(x)(y)(z)(Rxy → (Ryz → Rxz)) (*transitivity*: R is *transitive.*)
(x)(y)(¬x = y → (Rxy ∨ Ryx)) (*weak connectivity*: R is *weakly connected.*)

If R were to be interpreted as equality (=) on the set of natural numbers, we would expect to have as theorems of its fo theory, the following:

(x)Rxx (*reflexivity*: R is *reflexive.*)
(x)(y)(Rxy → Ryx) (*symmetry*: R is *symmetric.*)
(x)(y)(z)(Rxy → (Ryz → Rxz)) (*transitivity*:R is *transitive.*)
(x)(y)(z)(Rxy → (Rxz → Ryz)) (*the Euclidean property*: R is *Euclidean.*)

Quaere: Why should such a relation be called Euclidean?

Certain of the listed properties belong to groups of associated properties which are given systematic variations of a single label:

(x)Rxx (*reflexivity*)
¬(x)Rxx (*non-reflexivity*)
(x)¬Rxx (*irreflexivity*)
(x)(∃y)(Rxy → Rxx) (*weak reflexivity*)(sometimes called 'reflexivity' when the property represented by (x)Rxx is called 'total reflexivity'.)

Exercise 5.9

Which of the above are properties of *parenthood*?

(x)(y)(Rxy → Ryx) (*symmetry*)
¬(x)(y)(Rxy → Ryx) (*non-symmetry*)
(x)(y)(Rxy → ¬Ryx) (*asymmetry*)
(x)(y)(Rxy → (Ryx → x = y)) (*antisymmetry*)

Exercise 5.10

Which of the above are properties of *parenthood?*, of *brotherhood?*

$(x)(y)(z)(Rxy \rightarrow (Ryz \rightarrow Rxz))$ (*transitivity*)
$(x)(y)(z)(Rxy \rightarrow (Ryz \rightarrow \neg Rxz))$ (*intransitivity*)
$\neg(x)(y)(z)(Rxy \rightarrow (Ryz \rightarrow Rxz))$ (*non-transitivity*)

Exercise 5.11

Which of the above are properties of *parenthood*?, of *brotherhood*? of *half-sisterhood*?

5.7.1 Enthymemes

An *enthymeme* is an argument in which either the conclusion or a significant premiss is implicit. Frequently the implicit premiss is a sentence asserting of some relation occurring in the explicit premisses that it has a certain first-order property.

Example:

Sarah is the mother of Isaac; therefore Isaac is not the mother of Sarah.

The missing premiss is the sentence that if one thing is the mother of a second, then the second is not the mother of the first:

$(x)(y)(Rxy \rightarrow \neg Ryx)$

which asserts that *motherhood* is asymmetric. Since that wff would be a theorem in the first-order theory of motherhood, it may be introduced as a line of a proof by theorem introduction. Alternatively, it may be made explicit as a premiss of the argument if the proof to be produced is regarded as a proof in quantificational logic with identity, rather than as a proof in the first-order theory of *motherhood*. The argument would then be represented as:

$(x)(y)(Rxy \rightarrow \neg Ryx), Rmn \vdash \neg Rnm$

5.7.2 Some additional notation

It is convenient on occasions to have notation that permits us to express operations on relations. One such operation is composition. We use the same notational device that we introduced in Chapter 2 for the composition of truth-functions. The relation

$$R \circ R'$$

is the composition of R and R'. It consists of the pairs a, b satisfying

$$(\exists x)(Rax \wedge R'xb)$$

As an example, the relation expressed by *is the grandmother of* is the composition of the relation *is a parent of* and the relation *is the mother of*. Evidently from this example, the operation of composition is not commutative. That is, in general, R∘R' is not the same relation as R'∘R. If we compose the relations of our example in the opposite order, the resulting relation is the one expressed by *is a maternal grandparent of*.

Another such operation is *exponentiation*. For example, we can write R^2ab to abbreviate $(\exists z)(Raz \wedge Rzb)$; we can write R^3ab to abbreviate $(\exists z)(\exists w)((Raz \wedge Rzw) \wedge Rwb)$ and so on. We give these an intuitive reading as 'a is two R-steps (three R-steps) from b' or 'a is two (three) steps of the relation R before b.' Within this convention, R^1ab is alternative notation for Rab; R^0ab expresses a = b; $R^{-1}ab$ expresses Rba, and thus R^{-1} is common notation for the converse relation of R. So for example if we represent 'a is an ancestor of b' by Aab, we express 'a is a descendent of b' by $A^{-1}ab$. Accordingly we can represent the symmetry of R as a relationship between R and its converse, R^{-1}:

$$(x)(y)(Rxy \rightarrow R^{-1}xy).$$

And of course the relation $(R^{-1})^{-1}$ is the relation R. The notation permits us also to represent relationships between relational properties. For example, we can see transitivity and denseness as representing converse properties. If we read 'Rab' as 'a is one step of the relation R before b', and '$(\exists z)(Raz \wedge Rza)$' as 'a is two steps of the relation R before b', then transitivity may be understood as saying that an individual two steps before another is also one step before it; by contrast denseness can be understood as saying that an individual one step before another is also two steps before it.

In L_q, we can also express relations among relations. As an example, the relation of *monotonicity* between R and R' In this case *R is monotonic along R'* is expressed by the wff

$$(x)(y)(z)(Rxy \wedge R'yz \rightarrow Rxz)$$

For example the relation expressed by 'is an ancestor of' is monotonic along the relation expressed by 'is a parent of', since if a is an ancestor of b and b is a parent of c, then a is an ancestor of c.

Some of the properties we have listed above can be understood in the language of exponentiation, composition and monotonicity. In exponential notation, transitivity is expressed as

$$(x)(y)(R^2xy \to R^1xy).$$

A relation is *dense* iff for every pair in the relation there is an object that comes 'after' the first of the pair and 'before' the second. Consider, for example, the set of fractions between 0 and 1 ordered by the *is less than* relation. Notice that for any pair of such fractions, there is a fraction that is greater than the first and smaller than the second. Denseness of a relation R is expressable by

$$(x)(y)(Rxy \to ((\exists z)(Rxz \wedge Rzy))).$$

But it is also expressable using exponential notation as

$$(x)(y)(R^1xy \to R^2xy).$$

And we can also express these properties using the language of composition. A relation, R is transitive iff for every n, R is monotonic along R^n. Alternatively, R is transitive if every pair in R∘R is also in R. A relation R is dense iff every pair in R is also in R∘R. Alternatively, R is dense iff for every n, R^n is monotonic along R.

We can also compare properties as to their inferential strength. To give an example, denseness is a property *weaker than* pointwise reflexivity, which also says that if a is one step before b, then a is also two steps before b, but specifies that the intermediate individual is b itself. The sentence expressing pointwise reflexivity is interderivable with

$$(x)(y)(Rxy \to (\exists z)((Rxz \wedge Rzy) \wedge R^0zy))).$$

Exercise 5.12

Demonstrate the last claim.

Exercise 5.13

Give examples of dense relations.

Exercise 5.14

Demonstrate that a relation R is antisymmetric if the identity relation is monotonic along $R \circ R^{-1}$.

Exercise 5.15

For a symmetric relation R, define a relation R′ such that R is euclidean if R is monotonic along R′. Construct a proof that demonstrates the correctness of your answer.

Exercise 5.16

What property asserts that points that are no steps apart are also one step apart? Demonstrate your answer.

Exercise 5.17

Identity is monotonic along what relation?

Exercise 5.18

Demonstrate that if a relation R is not monotonic along identity, then R is reflexive.

Exercise 5.19

Demonstrate that if a relation R is reflexive and Euclidean, then R is symmetric and transitive.

Proof: We provide proofs for the two sequents:

(a) $(x)Rxx, (x)(y)(z)(Rxy \land Rxz \rightarrow Ryz) \vdash (x)(y)(Rxy \rightarrow Ryx)$

(b) $(x)Rxx, (x)(y)(z)(Rxy \land Rxz \rightarrow Ryz)$
          $\vdash (x)(y)(z)(Rxy \land Ryz \rightarrow Rxz)$

(a)	1	(1)	$(x)Rxx$	A
	2	(2)	$(x)(y)(z)(Rxy \land Rxz \rightarrow Ryz)$	A
	3	(3)	Rab	A(CP)
	1	(4)	Raa	1,UE
	2	(5)	$(y)(z)(Ray \land Raz \rightarrow Ryz)$	2,UE

2	(6)	(z)(Rab ∧ Raz → Rbz)	5,UE
2	(7)	Rab ∧ Raa → Rba	6,UE
1,3	(8)	Rab ∧ Raa	3,4,∧I
1,2,3	(9)	Rba	7,8,MPP
1,2	(10)	Rab → Rba	3,9,CP
1,2	(11)	(y)(Ray → Rya)	10,UI
1,2	(12)	(x)(y)(Rxy → Ryx)	11,UI

(b)				
	1	(1)	(x)Rxx	A
	2	(2)	(x)(y)(z)(Rxy ∧ Rxz → Ryz)	A
	1,2	(3)	(x)(y)(Rxy → Ryx)	1,SI(a) *supra*
	4	(4)	Rab ∧ Rbc	A(CP)
	1,2	(5)	(y)(Ray → Rya)	3,UE
	1,2	(6)	Rab → Rba	5,UE
	4	(7)	Rab	4,∧E
	1,2,4	(8)	Rba	6,7,MPP
	1	(9)	Rbb	1,UE
	2	(10)	(y)(z)(Rby ∧ Rbz → Ryz)	2,UE
	2	(11)	(z)(Rba ∧ Rbz → Raz)	10,UE
	2	(12)	Rba ∧ Rbc → Rac	11,UE
	4	(13)	Rbc	4,∧E
	1,2,4	(14)	Rba ∧ Rbc	8,13,∧I
	1,2,4	(15)	Rac	12,14,MPP
	1,2	(16)	Rab ∧ Rbc → Rac	4,15,CP
	1,2	(17)	(z)(Rab ∧ Rbz → Raz)	16,UI
	1,2	(18)	(y)(z)(Ray ∧ Ryz → Raz)	17,UI
	1,2	(19)	(x)(y)(z)(Ray ∧ Ryz → Raz)	18,UI

Exercise 5.20

Demonstrate that if a relation R is symmetric and transitive, then R is Euclidean.

Proof: We provide a proof for the sequent,

(c)   (x)(y)(Rxy → Ryx), (x)(y)(z)(Rxy ∧ Ryz → Rxz)
                    ⊢ (x)(y)(z)(Rxy ∧ Rxz → Ryz)

1	(1)	(x)(y)(Rxy → Ryx)	A
2	(2)	(x)(y)(z)(Rxy ∧ Ryz → Rxz)	A

3	(3)	Rab $\wedge$ Rac	A(CP)
3	(4)	Rab	3,$\wedge$E
1	(5)	(y)(Ray $\rightarrow$ Rya)	1,UE
1	(6)	Rab $\rightarrow$ Rba	5,UE
1,3	(7)	Rba	4,6,MPP
3	(8)	Rac	3,$\wedge$E
1,3	(9)	Rba $\wedge$ Rac	7,8,$\wedge$I
2	(10)	(y)(z)(Rby $\wedge$ Ryz $\rightarrow$ Rbz)	2,UE
2	(11)	(z)(Rba $\wedge$ Raz $\rightarrow$ Rbz)	10,UE
2	(12)	Rba $\wedge$ Rac $\rightarrow$ Rbc	11,UE
1,2,3	(13)	Rbc	9,12,MPP
1,2	(14)	Rab $\wedge$ Rac $\rightarrow$ Rbc	3,13,CP
1,2	(15)	(z)(Rab $\wedge$ Raz $\rightarrow$ Rbz)	14,UI
1,2	(16)	(y)(z)(Ray $\wedge$ Raz $\rightarrow$ Ryz)	15,UI
1,2	(17)	(x)(y)(z)(Rxy $\wedge$ Rxz $\rightarrow$ Ryz)	16,UI

5.7.3 n-ary relations and their properties

All of what has been said above with regard to binary relations can be taken to apply *mutatis mutandis* to ternary, quaternary, and in general to n-ary relations for any n. However, the conventions governing the naming of first-order relational properties are not so well established in the case of relations of *arity*[4] greater than 2. For example, a symmetric binary relation is one satisfying the condition

$$(x)(y)(Rxy \rightarrow Ryx)$$

but that fact dictates no single way of picking out the class of ternary or quaternary relations that is suitably labelled 'symmetric'. How we generalize the label for relations of higher arity will depend upon how the property is described, perhaps even how we describe the syntactic relationship between Rxy and Ryx. If we think of the second as reversing the first two variables of Rxy, then we might expect a symmetric ternary relation to satisfy the condition:

$$(x)(y)(z)(Rxyz \rightarrow Ryxz).$$

---

4. An n-ary relation is thought of as a set of n-tuples of individuals. The words *arity* and *adicity* provide a convenient informal way of referring to the size of the n-tuples of objects in the relation. Thus a binary relation (a set of ordered pairs) has arity (or adicity) 2; a ternary relation (a set of ordered triples) arity 3, and so on.

If we think of the second as reversing the last two variables of Rxy, then we might expect a symmetric ternary relation to satisfy the condition

$$(x)(y)(z)(Rxyz \rightarrow Rxzy).$$

Alternatively, if we think of the relationship cyclically, then we might expect a symmetric ternary relation to satisfy the condition

$$(x)(y)(z)(Rxyz \rightarrow Rzxy).$$

Coming at the problem from a specific theoretical or practical interest, we might find some quite distinct generalization more natural because it is urged by important features of the application. In logic, one important application of first-order theories of relations is in the study of modal logics, logics of various kinds of necessity ($\Box$A is read 'Necessarily A'.) One semantic representation of modal logics gives truth-conditions for such sentences on relational structures called *frames*. In the standard binary idiom $\Box$A is said to be true at one point x in such a structure if and only if A is true at every point y to which x is in the relation R of the structure. It is seen that under certain first-order assumptions about R, certain formulae of the language are true everywhere. For example, in structures having a symmetric R, the wff (called the Brouwersche formula)

$$P \rightarrow \Box\neg\Box\neg P$$

is true everywhere. So in the standard generalization of these structures to those in which, for some n, an n-ary R replaces the binary one, it is natural to apply the label *symmetry* to whatever condition the n-ary relation must satisfy in order to make the Brouwersche formula true everywhere. As it happens this condition is unlike any of the possible generalizations already listed, though it is still intuitively thought of as a kind of symmetry. For ternary R, this is the condition

$$(x)(y)(z)(Rxyz \rightarrow Rzxx \lor Ryxx).$$

For quaternary R, it is the condition

$$(x)(y)(z)(w)(Rxyzw \rightarrow (Ryxxx \lor Rzxxx) \lor Rwxxx)$$

and so on. All of the following conditions, given here just for the ternary case, are the standard generalizations of the binary conditions bearing the same name.
Reflexivity

$$(x)Rxxx$$

Seriality

$$(x)(\exists y)(\exists z)Rxyz$$

Transitivity

$$(x)(y)(y')(z)(z')(w)(w')(Rxyy' \wedge (Ryzz' \wedge Ry'ww') \rightarrow (Rxzz' \vee (Rxww'$$
$$\vee (Rxzw \vee (Rxzw' \vee (Rxz'w \vee (Rxz'w')))))))$$

The Euclidean property

$$(x)(y)(y')(z)(z')(Rxyy' \wedge Rxzz' \rightarrow Rzyy' \vee Rz'yy')$$

Quasi-binarity

$$(x)(y)(z)(Rxyz \rightarrow Rxyy \vee Rxzz)$$

If a ternary relation R has the last listed property, then the same wffs are made valid as would be made valid in a structure with a binary relation. R is therefore called *quasi-binary*.

Exercise 5.21

Use **Simon**'s proof editor to construct proofs for the following sequents using only primitive rules and *modi*.

(a) $(x)(y)(z)(Rxyz \rightarrow Rzxy) \vdash (x)(y)(z)(Rxyz \rightarrow Ryzx)$

(b) $(x)(y)(\exists z)Bxyz, (x)(y)(z)(Bxyz \rightarrow Bxzy)$
$\vdash (x)(y)(\exists z)(\exists w)(Bxyz \wedge Bxzw)$

(c) $(x)((\exists y)Ryx \rightarrow (z)Rxz) \vdash (y)(z)(Ryz \rightarrow Rzy)$

Exercise 5.22

Demonstrate that if ternary R is reflexive, then R is serial. (For this and succeeding exercises use **Simon**'s proof editor to construct the required proofs.

Exercise 5.23

Demonstrate that if ternary R is Euclidean and reflexive, then R is quasi-binary.

Exercise 5.24

Demonstrate that if ternary R is symmetric, then R is quasi-binary.

Exercise 5.25

Demonstrate that if binary R is serial, transitive and irreflexive, then R is asymmetric.

5.8 Experimenting With First-order Theories

The material throughout this section is intended to be an adventure, and should be approached in that spirit. Some of its exercises can be submitted and graded, but for some of the exercises, the outcome is entirely up to the reader. You will be asked to formulate some definitions, and the syntax in which they are formulated will be carefully prescribed. But the definitions themselves will be the creation of the reader. Of course we are dependent here upon some family of common notions about the items we will be defining, and the definitions will have to be such that they enable the reader to construct some proofs in later exercises. But they are also experimental. If the definitions permit implausible proofs, the definitions will require revision, but there is no requirement nor even an expectation of convergence upon some correct set of definitions. Simply, what definitions we accept will determine which proofs we can construct. The interplay between the two activities, defining and proving, gives the section its point. Of course it is hoped that the subject matter will be of more than idle interest.

5.8.1 Family, foto

We have briefly explored the idea of a first-order theory, particularly through our introduction to the first-order theory of identity, and our subsequent discussion of first-order theories of binary and in general n-ary relations. Here we provide a little laboratory in which to try out what we have learned in the development of a first-order theory of biological affiliation. The aim here is to codify, that is, to reduce to a small number of fundamental principles the formal structure of a mammalian species, such as that of human primates. This sounds very grand, but in fact it must reduce to a few very familiar primitive relationships, the properties of which are easily given. When all of the serious, academic remarks have been made, and all of the possible complications and counterexamples set aside, we will find ourselves considering some literally familiar notions: those connected with the biological structure of human family relationships. Hence the heading "Family, foto (first-order theory of)". First the stipulations and disclaimers.

Of course, any biological species considered over a sufficient period of time must be a non-classical set. Every human has non-human ancestors, and between the clearly non-human ancestors and the clearly human ones, there must be ancestors whose status is unclear vis-à-vis humanity. No purely formal theory will enable us to decide membership in the set. So to say that we can lay bare *the* formal structure is perhaps unwarrantedly optimistic. Nothing that we do in a first-order language will equip us to offer a *biologically* adequate definition of a first human. The best we could do would be to satisfy one or another mythology by postulating humans without biological parents. And again, the species is a biological item of which the regularities must be merely statistical. There is no such thing as a normal human being. All that can be normal is a distribution in a population. So any set of fundamental principles must be in some measure simplifying assumptions. We will not try on our first attempt, to take into account such phenomena as hermaphroditism, parthenogenesis, conjoined twins, or cloning, nor will we have representations of gender differences that are irrelevant to reproduction. From a sociological or sexological point of view our theory will therefore be rather bland. We cannot hope to represent much more than the familiar (and less familiar) relationships that sexual generation creates.

A second stipulation should be made explicit. The relationships about which we will theorize are biological ones, not civil. That is to say, there is no place in our theory for step parents, for adoptive parents or siblings, nor for step sisters, step brothers or step parents. When we speak of aunts and uncles, we mean biological aunts and uncles, not those that one acquires by one's own marriage or by the marriage of a biological uncle or aunt.

Once again, we must say a word or two about the formal status of such a theory. We will study this fo theory with its intended use in mind, but what we are labelling its intended use can be regarded as merely providing a *reading* for the elements of its language; it need not be regarded as the sole available *interpretation* of it. With no alteration at all to its special principles or its definitions, it will be available for many distinct classificatory applications, and with modifications many more. Much of the point of acquiring facility with such a theory is to question the efficiency of its rules, and to explore the consequences of altering its foundational principles upon available interpretations and applications.

5.8.2 The language

To the stock of logical constants of a first-order theory of identity, we add four new predicate symbols, $<$, $\mathcal{F}$, $\mathcal{M}$, $\mathcal{P}$ with corresponding clauses to the definition of Φ.

    1.   If t and t' are terms, then $t < t'$ is an atom.

    2.   If t is a term, then $\mathcal{F}t$ is an atom.

    3.   If t is a term, then $\mathcal{M}t$ is an atom.

4.   If t and t′ are terms, then $\mathcal{P}tt′$ is an atom.

which establish that $\mathcal{F}$ and $\mathcal{M}$ are 1-place, and $<$ and $\mathcal{P}$ 2-place predicates. We also establish that $<$ takes the infix position.

5.8.3 The principles

If t and t′ are terms, then any of the following may be introduced as a line of a proof with no required assumptions.

[Irr] $\neg\, t < t$

[Trans] $t < t′ \wedge t′ < t″ \rightarrow t < t″$

[Ser] $(\exists y)\, t < y$

[G] $\neg\,\mathcal{F}t \vee \neg\,\mathcal{M}t$

[Sub] $\mathcal{P}tt′ \rightarrow t < t′$

[FP] $(\exists x)((\mathcal{F}x \wedge \mathcal{P}xt) \wedge (y)(\mathcal{F}y \wedge \mathcal{P}yt \rightarrow x = y))$

[MP] $(\exists x)((\mathcal{M}x \wedge \mathcal{P}xt) \wedge (y)(\mathcal{M}y \wedge \mathcal{P}yt \rightarrow x = y)).$

We read 't < t′' as 't antecedes (is born earlier than) t′'. It will already be evident that our reading of '$\mathcal{F}$t'is 't is female', of '$\mathcal{M}$t', 't is male', and of $\mathcal{P}tt′$, 't is a parent of t′. On this reading, [Irr], [Trans] and [Ser] will impose on our domain the irreflexivity, transitivity, and seriality of antecedence. [G] will impose the condition that no object in the domain is both male and female; [FP] that every object in the domain has exactly one female parent, and [MP] that every object in the domain has exactly one male parent. [Sub] is a bridge principle that tells us that parenthood is a subrelation of antecedence, that is, that all parents are antecedents of their offspring. The proof that functional reproductive bisexuality is ruled out by [G] is the following:

113 $\vdash (x)(\neg\,\mathcal{F}x \vee \neg\,\mathcal{M}x)$

        (1) $\neg\,\mathcal{F}a \vee \neg\,\mathcal{M}a$ G

        (2) $(x)(\neg\,\mathcal{F}x \vee \neg\,\mathcal{M}x)$ 1,UI

The proofs of the following are left as exercises.

114 $\vdash (x)\neg\, x < x$

115   ⊢ (x)(y)(z)(x < y ∧ y < z → x < z)

116   ⊢ (x)(∃y)y < x

117   ⊢ (x)¬Pxx

118   ⊢ (x)(y)(Pxy → ¬Pyx)

119   ⊢ (z)(∃x)((ℱx ∧ 𝒫xz) ∧ (y)(ℱy ∧ 𝒫yz → x = y))

120   ⊢ (z)(∃x)((ℳx ∧ 𝒫xz) ∧ (y)(ℳy ∧ 𝒫yz → x = y))

It should be clear that in articulating those principles, we are not making a universal claim about biological nature or even mammalian or human reality. We are merely restricting the domain of objects about which we propose to theorize to the domain of objects for which the implicit claims are true. Any theorems that we prove will have application only to objects in such a domain. Whether the fo theory is practically useful might depend upon whether there are an interesting body of objects in the world worth studying in this way that such a domain could be taken to represent. Our interests here are paedogogical; it is for the reader to decide whether the exercise is a paedogogically useful one. One early question in the development of such a theory regards the choice of principles. One might think that [FP] and [MP] represent reasonable restrictions, but query the choice of [G] over other candidates that might have suggested themselves. This gives us our first exercise.

Exercise 5.26

What would be the effect of replacing [G] with

[G′]   ℱt ∨ ℳt

That is, what non-primitive theorems could be proved in the resulting fo theory that would distinguish it from the one we are proposing?

A second question, to which we must eventually give a formally correct answer, concerns the expressive power of the chosen language within the area of application of the theory. This provides our second exercise, which might better be described as a commission, as its execution requires decisions rather than correct answers. We can illustrate the problem by contrasting a sister with a half-sister. Question; is a half-sister a sister? The answer is not dictated by the form of words. A jailhouse lawyer need not be a lawyer; artificial vanilla is not vanilla; fool's gold is not gold; by contrast artificial insemination is a form of insemination. In any case there is some latitude in the way that we answer the question. The distinction between sister and half-sister is not jeopardized whether we think of a sister of Alison as a female sharing at least one parent with Alison or as a female sharing two parents with her. For both are distinct from a female sharing exactly one parent with Alison. Parallel remarks could be made about brothers and siblings. We may use the language as we like within limitations, but the use will affect other definitions such as those of *aunt* and *uncle*, and they in turn may affect the definition of *cousin*. Again the definitions that we provide will affect the availability of proofs, and therefore the status of certain natural language claims, which on one definition might be represented by wffs that are theorems and by another definition not. We must think of the task therefore as a kind of experiment. In the first instance, we propose reasonable-seeming definitions, but later we can revisit and revise. At any rate the definitions that we now propose experimentally will become the definitions that we will invoke in later proofs.

Exercise 5.27

Define each of the following predicates in such a way that it will bear the reading proposed for it. Use earlier definitions as components of later ones if you wish.

Df.S	Stt' (t is a sister of t')
Df.B	Btt' (t is a brother of t')
Df.S'	$S'tt'$ (t is a sibling of t')
Df.S''	$S''tt'$ (t is a half-sibling of t')
Df.M'	$M'tt'$ (t is a mother of t')
Df.V	Vtt' (t is a father of t')
Df.A	Att' (t is an aunt of t')
Df.U	Utt' (t is an uncle of t')

Df.C Ctt' (t is a cousin of t')
Df.G Gtt' (t is a grandparent of t')
Df.G' $G'tt'$ (t is a greatgrandparent of t')

A third question concerns the theorems involving primitive and defined predicates that represent fo properties of the corresponding relations. We demonstrate that a relation has a fo property by demonstrating that the corresponding wff is a theorem. We demonstrate that it does not, by exhibiting a partial model which is consistent with the primitive principles and definitions, but in which the corresponding fo wff fails.

Exercise 5.28

For each of the following, exhibit a proof if it is correct or exhibit a falsifying partial model if it is not.

(a)   Offspringhood is serial.
(b)   Parenthood is intransitive.
(c)   Antecedence is asymmetric.
(d)   Parenthood is asymmetric.
(e)   Parenthood is irreflexive.
(f)   Siblinghood is symmetric.
(g)   Cousinhood is symmetric.
(h)   Brotherhood is transitive.

A fourth question might arise concerning the need for the antecedence relation in such a theory. To be sure, the asymmetry and transitivity of < enables us, as the previous exercise reveals, to demonstrate the asymmetry and irreflexivity of P through the use of the monotonicity principle [Sub], but we could have imposed those conditions by appropriate principles for P, and without mentioning <. Such a move would render the remaining < principles otiose. That is the subject of the next exercise.

Exercise 5.29

Replace [Sub] with two principles that impose on the parenthood relation asymmetry and irreflexivity. Exhibit (a) a triple and (b) a 4-tuple which satisfy the resulting set of principles for P but which are counterintuitive for its intended interpretation.

5.8.4 Biology is not first-order

Anyone in the grip of the belief that we are all descended from an original human couple will wish to reject the principles [FP] and [MP] in favour of a principle that dictates the existence of exactly two objects in the domain that are parents, but themselves have none. Such a fo theory would have the advantage that its principles could be satisfied in a finite domain of individuals. But we might share with such a person the view that the intended interpretation of the theory we are experimenting with demands that its principles be satisfied in a finite domain, even if we don't share the more radical theory of first human ancestors. On an evolutionary understanding nothing can non-arbitrarily bear the label 'first human female' or 'first human male'. On either account we must reject the principles of the fo theory under study. However the biological realm might have come into being, no one supposes that it did not do so. So adherents to the opposing views have it in common that they both suppose that everything biological has non-biological antecedents. In either case, finitude figures in our understanding of the biological. However, elements of the two previous exercises will enable us to complete another.

Exercise 5.30

> Demonstrate (by mathematical induction) that the fo theory we are considering has no finite models.

In fact there is no fo sentence we can add to our principles that will impose such finitude, since finitude is not a first-order property. Any such theory that has a model has a model on the set of natural numbers.

Does this render such a theory useless? The answer is that in that one respect in which it fails, every fo theory must fail. But that does not render it useless, for its fault is not in misrepresenting the local properties of biological affiliations, but in representing the structure of them as always having been as it is. The intellectual challenge of evolutionary biology lies in divining from temporally local observations of change, the principles that govern the evolution of those changes themselves. The character of evolutionary change has itself evolved and is certainly changing. To provide a concrete example, we have no idea what the ancestry of sexual reproduction was in sufficiently early stages of biological evolution; it may have its origins in highly particular forms of partial replication by pairs of RNA molecules in the pre-biotic period. There may be no way of extending the language of sexual difference to such complementary molecules. No doubt some features of xP would have application, perhaps [Mon], but others certainly would not.

For purposes of exploring the construction of fo theories, we have simply given ourselves vastly more temporal work-space than we require for anything we will try to do.

Exercise 5.31

Formulate each of the following claims as a wff in the language of this fo theory, construing each in such a way that there is a proof demonstrating its theoremhood. For each, construct such a proof.

(a)   Everyone has exactly two parents.
(b)   Everyone has at most two grandfathers.
(c)   Cousins who are siblings are half-siblings.

Exercise 5.32

Construct a partial model which illustrates the affiliation between each of the following pairs.

(a)   Cousins with a common sibling
(b)   Second cousins
(c)   First cousins once removed

Exercise 5.33

Assess the following. Construct a proof that demonstrates the claim, or construct a partial model to show the claim to be falsifiable.

An aunt of an aunt is a grand aunt.[5]

Exercise 5.34

For each of the following arguments, represent it as a sequent in the fo theory of affiliations using the lower case initials of the natural proper names as their formal counterpart. Determine whether the sequent is valid or invalid in your fo theory. If it is invalid, exhibit a partial model in which it fails; if it is valid, construct a proof.

(a)   Ken has at least one sibling.

(b)   Ken has no siblings.

(c)   All of Karen's siblings have dimples. Therefore Karen has dimples. (Dx: x has dimples)

(d)   Some of Mabel's sisters play chess. In fact Mabel alone among Naomi's daughters does not. Therefore Mabel plays chess. (Px: x plays chess)

(e)   At most one of Karl's siblings has children. Karl's son, Matt has a cousin. Therefore, exactly one of Karl's siblings has children.

(f)   Noah's mother is Lillian's mother's only daughter. Therefore Noah is Lillian's son.

(g)   Mark's father is Noah's mother's only son. Therefore, Noah is Mark's aunt.

(h)   All of Lucas's cousins are cousins of Martyn. Therefore Lucas and Martyn are siblings.

---

5. Here and throughout we speak (correctly as it happens) of grand uncles and grand aunts, not great uncles and great aunts. Grand aunts and grand uncles bear the relationships to grandparents that uncles and aunts bear to parents. Likewise great grand aunts are aunts of grandparents and so on.

(i)   Lana has some cousins that are cousins of Mary, and none
      that are not. Therefore either Lana and Mary are siblings
      or cousins.

(j)   Matt and Nan are siblings. Therefore all of Matt's cousins
      are cousins of Nan.

(k)   Matt and Nancy are cousins. Therefore, all of Mat's
      siblings are cousins of Nancy.

(l)   Matt and Nancy are cousins. Therefore, all of Mat's
      cousins are cousins of Nancy.

(m)   Larry and Norris are cousins. Therefore, Larry's father is
      Norris's uncle.

(n)   Larry's father is Norris's uncle. Therefore, Larry and
      Norris are cousins.

(o)   Lisa and Margaret are cousins. Gillian's father is not
      Margaret's uncle. Margaret's mother is not Lisa's aunt.
      Therefore Gillian's mother and Margaret's father are
      siblings.

(p)   Marni and Nathalie are sisters. So every grandfather of
      Marni's is a grandfather of Nathalie's.

(q)   Lucas and Martyn are brothers. Therefore they have at
      least one grandmother in common.

(r)   Anyone who is ambitious gets it from a parent. Therefore
      if Morris is ambitious, so must be one of his grandparents.

(s)   Matt is ambitious. Anyone who is ambitious gets it from a
      parent. Neither of Matt's paternal grandparents was ever
      ambitious. Therefore, Matt's mother is ambitious. (Ax: x is
      ambitious)

(t)   One has blue eyes if and only if both of one's parents had
      blue eyes. Therefore if Matthew has blues eyes, so do all
      of his uncles. (Bx: x has blue eyes)

(u)   One is fair if and only if both of one's parents are fair. All
      of Martyn's siblings are swarthy, though Martyn, like his
      mother, is fair. So Martyn's father is not fair. (Fx: x is fair)

5.9 Quantificational Models Revisited

5.9.1 Introduction

We have said that the universal quantifier is to be understood as a generalized conjunctive connective, and the existential quantifier as a generalized disjunctive one. We have, moreover, used these interpretions in domains of crucial size to test the validity of arguments expressible in the monadic fragment of L_q. The following sections will take us a little beyond those concrete and simple understandings to a more general semantic account of the quantifiers, one which confirms our reading of (∃x) as *for some x* and (x) as *for all x*. This account will be sufficiently general as to enable us to prove the soundness of L_q. It will also be sufficiently specific that it would enable us to prove completeness.

We understand propositional wffs abstractly as being about states of affairs, and so, somewhat less abstractly, we also understand quantificational wffs. Less abstractly because, in the quantificational case, we acknowledge that one component of a description of a state of affairs will be an account of the objects present in it and their properties and relationships. Both are necessary ingredients of an account of truth for the atomic sentences of the language, since every such sentence consists of a predicate symbol followed by some finite number of terms. We will want to say that such a sentence is true if and only if the relation that the predicate term is taken to stand for holds among the objects that the terms are taken to name. We will want to say that Fm is true iff the property that F is taken to pick out holds of the object that the proper name m is taken to name. If F is understood as the property of being fickle, and m is taken to name Mike, then on that understanding of F and m, Fm will be true iff Mike is fickle. Similarly for relations. If G is interpreted as the relation of *being more gregarious*, m is taken to refer to Mike and n is taken to refer to Neil, then Fmn will be true if Mike is more gregarious than Neil, and otherwise false. Propositional logic in some philosophical settings is referred to as the logic of unanalysed propositions; by contrast, quantificational logic is referred to as the logic of analysed propositions. The distinction of labels corresponds to a difference in the character of propositional and quantificational atoms.

Philosophically the most that can be said of propositional atoms is that they are to be understood as the sorts of things that can take truth values. Quantificational atoms also take truth values, but their truth values are dependent upon the interpretation of their components. Thus, while the basis of a propositional model is simply an assignment of truth values to atoms, the basis for a quantificational model is the fixing of the reference of proper names and the interpretation of predicate symbols. It is only when we know what "m" names, and what property "F" is taken to represent that we can answer the question whether the atom Fm is true or false. Informally, Fm is true if the object named by m has the property that F is interpreted as; it is false otherwise. But m can be made to name any object, and F can be taken to represent any property: hence the multiplicity of models on a single domain of individuals. Different models may take F to represent different properties, and m to name different objects. To sum up then, we will interpret proper names as names of objects in a domain; we will interpret predicate symbols as properties; and we will allow the interpretation to dictate the truth value of the atoms in which only proper names occur. Atoms containing occurrences of arbitrary names will receive slightly different treatment, and we will have to build an account of truth for all other connectives upon the those for atoms, just as we did for propositional logic. As in the case of propositional logic, we will be particularly interested in the wffs that will be true regardless of what objects proper names name and regardless of what properties the predicate symbols are taken to represent. These wffs we will take to be valid. As in propositional logic, we want to have only valid wffs as theorems (the weak soundness requirement), and we want to have all valid wffs as theorems (the weak completeness requirement).

5.10 Quantificational models: the details

5.10.1 The truth- and validity-conditions of atoms

A quantificational model $\mathcal{M}$ is an ordered pair $\langle \mathcal{D}, \mathcal{I} \rangle$, where $\mathcal{D}$ is a non-empty set called the *domain* of the model and $\mathcal{I}$ is an interpretation. We describe these in turn. To say that the domain is non-empty is of course to say that it has at least one element. That is the *only* restriction on $\mathscr{D}$. So if you wish to think concretely about a particular domain of a particular model, just choose any non-empty set that comes familiarly to mind. It is natural for a logician to think of the set of natural numbers, but if you wish to think about the set of heads of state of your country, or the set of its states or provinces or counties or cities, or the set of buttons down the front of your shirt, or the set of hairs on your head, that will do nicely, so long as your shirt has at least one button down the front or, in the second case, you are not completely bald. Any non-empty set can be the domain of a model.

An interpretation, $\mathscr{I}$ (script 'I') can be thought of as fixing the reference of every proper name in the language. Officially it assigns every proper name to an object in the domain. If the domain is finite, then evidently some objects in it must have infinitely many proper names assigned to it; if the domain is an infinite set such as the set of natural numbers, then each object can have a unique proper name, but there will be uncountably[6] many ways of assigning proper names to objects, and so uncountably many distinct models having the set of natural numbers as their domain. The interpretation also fixes "meanings" of predicate symbols. Since an atom can consist of any predicate symbol preceding any finite number (zero or greater) of proper names, $\mathscr{I}$ must give to each predicate symbol P a collection $\mathscr{I}_n(P)$ of ordered n-tuples, for every n equal to or greater than zero. Mathematically there is exactly one 0-tuple, so we can understand the assignment of a set of zero-tuples as the assignment of a truth-value, 0 if the set of 0-tuples is empty, 1 otherwise. This observation brings us to the subject of truth-conditions for this restricted class of atomic wffs: those consisting of a predicate letter followed by zero or more occurrences of proper names.

We write

$$\mathscr{M} \models A$$

to be read "The wff A is *valid* in the model $\mathscr{M}$" or "The model $\mathscr{M}$ *validates* A" and

---

6. To say that the set of assignments is uncountable is to say that there are more assignments than there are natural numbers to count them with.

$$\mathcal{M} \not\models A$$

to be read "The wff A is not valid in the model $\mathcal{M}$ or "The model $\mathcal{M}$ does not validate A".

Now let $\mathcal{M} = <\mathcal{D}, \mathcal{T}>$ be any model, let $p_1, \ldots, p_i \ldots$ be metalogical variables ranging over the proper names of L_q, and let P be any predicate symbol. Then we may say: For any n equal to or greater than 0,

$\mathcal{M} \models Pp_1 \ldots p_n$ if $<\mathcal{T}(p_1), \ldots, \mathcal{T}(p_n)>$ is a member of $\mathcal{T}_n(P)^7$; else $\mathcal{M} \not\models Pp_1 \ldots p_n$.

Evidently we have not yet given a complete account of the truth-conditions of atoms, since an atom can consist of a predicate symbol followed by any finite number ot *terms*. And although the category of terms comprises both proper and arbitrary names, we have so far considered only those comprising predicate symbols and proper names.

In the quantificational models we are describing, an arbitrary name will be treated as a special kind of individual variable, reflecting its special use in quantificational proofs. So the extension of our truth-theory to atoms involving arbitrary names is included in the extension to what we will call atomic *propositional functions*. A propositional function is an atomic propositional function in variables $v_1, \ldots, v_n$ iff the result of replacing all occurrences of $v_1, \ldots, v_n$ by occurrences of terms is an atom. Thus, for example, the propositional function in x and y

Fxamnby

is an atomic propositional function because the replacement of "x" and "y" by, say, "b" and "c" produces the atomic wff

Fbamnbc.

Now x and y are variables; they behave like pronouns "he", "it", "she", "her", "him". Propositional functions in which they occur depend for

---

7. We understand an n-ary relation as a collection of ordered n-tuples. For example, a binary relation is a set of pairs. In this *extensional* idiom, to say that a pair <Fred, Neil> is in the relation *taller than* is to say that Fred is taller than Neil.

their meanings upon what individual they refer to, and that can vary. This variabilility of reference is reflected in an additional requirement for the attribution of truth to atomic propositional functions. A model can give truth to an atomic propositional function only relatively to an assignment of references to individual variables.

An assignment λ (μ, ν, etc.)[8] assigns every individual variable, v and every term, t to an individual $\lambda(v)$, respectively $\lambda(t)$, in the domain, $\mathscr{D}$ of the model. If t is a proper name then $\lambda(t) = \mathscr{T}(t)$. We write

$\mathscr{M},\lambda \vDash Pt_1 \ldots t_n$ if $<\lambda(t_1), \ldots ,\lambda(t_n)>$ is a member of $\mathscr{T}_n(P)$; else $\mathscr{M},\lambda \nvDash Pt_1 \ldots t_n$.

In the former case we say that the wff is *satisfied* in the model $\mathscr{M}$ at the assignment λ, in the latter that the wff is *not satisfied* in the model $\mathscr{M}$ at the assignment λ.

A wff is valid in the model $\mathscr{M}$ iff it is satisfied in $\mathscr{M}$ at every assignment.

Clearly, if $Pt_1 \ldots t_n$ contains no occurrences of individual variables or arbitrary names, then if $\mathscr{M},\lambda \vDash Pt_1 \ldots t_n$, then $\mathscr{M} \vDash Pt_1 \ldots t_n$. That is, an atomic wff (or atomic propositional function in zero variables) is valid in model $\mathscr{M}$ iff it is satisfied in $\mathscr{M}$ at at least one assignment λ. (The explanation is that the truth-conditions of such atoms depend only upon $\mathscr{T}$, which is constant across assignments.)

5.10.2 Truth and validity more generally

More generally, the satisfaction relation $\vDash$ gives us a hierarchy of three notions

    (a) $\mathscr{M},\lambda \vDash A$ (A is satisfied in the model $\mathscr{M}$ at the assignment λ.)

---

8. These are the lower-case Greek letters *lambda*, *mu*, and *nu* respectively.

(b) $\mathcal{M} \models A$ (A is satisfied in the model $\mathcal{M}$ at every assignment λ, alternatively, A is valid in $\mathcal{M}$.)

(c) $\models A$ (A is valid in every model $\mathcal{M}$, alternatively, A is universally valid.)

Derivatively, we define quantificational entailment between an ensemble Σ and a wff A ($\Sigma \models A$) by:

$\Sigma \models A$ (Σ entails A) iff for every quantificational model $\mathcal{M}$ and every assignment λ, if $\mathcal{M},\lambda \models \sigma$ for every σ in Σ, then $\mathcal{M},\lambda \models A$.

There remains to provide the truth-conditions for the non-atomic propositional functions of Φ_q.

Propositional connectives

Truth-conditions for a propositional function in which the connective of longest scope is a propositional connective match exactly the truth-conditions given by the truth-table of the connective.

(a)   If A is of the form $\neg B$, then $\mathcal{M},\lambda \models A$ if $\mathcal{M},\lambda \not\models B$; else $\mathcal{M},\lambda \not\models A$.

(b)   If A is of the form $B \rightarrow C$, then $\mathcal{M},\lambda \models A$ if $\mathcal{M},\lambda \not\models B$ or $\mathcal{M},\lambda \models C$; else $\mathcal{M},\lambda \not\models A$.

(c)   If A is of the form $B \wedge C$, then $\mathcal{M},\lambda \models A$ if $\mathcal{M},\lambda \models B$ and $\mathcal{M},\lambda \models C$; else $\mathcal{M},\lambda \not\models A$.

(d)   If A is of the form $B \vee C$, then $\mathcal{M},\lambda \models A$ if $\mathcal{M},\lambda \models B$ or $\mathcal{M},\lambda \models C$; else $\mathcal{M},\lambda \not\models A$.

Quantifiers

Truth-conditions for quantified wffs require a little care. We want the quantifiers to satisfy their intended interpretations inasmuch as we want universally quantified wffs to "say" that what follows the quantifier within its scope is true of every object in the domain, and we want existentially quantified wffs to "say" that what follows the quantifier within its scope is true of at least one object in the domain. Our problem lies in saying that clearly and precisely. Some cases, such as (x)Fx will be unproblematically true when every object in the domain has property $\mathscr{I}$(F), and false otherwise. If such a wff is true at one assignment, it will be true at all assignments and therefore valid in the model. But what of such a case as (x)Fxab. The difficulty here is that at an assignment λ, this must "say" that every object in the domain is in the relation $\mathscr{I}_3$(F) with the object λ(a) and λ(b), the objects to which λ assigns the arbitrary names a and b.

Our truth-conditions cope with such complications by defining a ternary relation R, called an *alternativeness relation* as follows:

Let $\mathscr{M}$ = <$\mathscr{D}$, $\mathscr{I}$> be a model and λ and μ assignments of individual variables and arbitrary names to $\mathscr{D}$ and v an individual variable. We say that $R_v\lambda\mu$ (μ is a v-alternative to λ) iff for every individual variable v' other than v, and for every arbitrary name a, λ(v') = μ(v') and λ(a) = μ(a). Thus an x-alternative to an assignment λ is an assignment μ that is identical to λ on all assignments other that its assignment of x. We record this as $R_x\lambda\mu$.

Notice that for any variable v, R_v is an equivalence relation, that is, R_v is reflexive, transitive and symmetric. The first of these properties should not be overlooked. Every assignment λ is identical to itself on all assignments of individual variables, and therefore identical to itself on all assignments other than that of v.

We can now take advantage of this device in truth-conditions for quantificational wffs.

Let $\mathscr{M}$ = <$\mathscr{D}$, $\mathscr{I}$> be a model and λ an assignment of individual variables and arbitrary names to $\mathscr{D}$. Let v be an individual variable and A(v) a propositional function in the variable v. Then

$\mathcal{M},\lambda \models (v)A(v)$ if for every assignment μ, if $R_v\lambda\mu$, then $\mathcal{M},\mu \models A(v)$; else $\mathcal{M},\lambda \not\models (v)A(v)$

$\mathcal{M},\lambda \models (\exists v)A(v)$ if for some assignment μ, $R_v\lambda\mu$ and $\mathcal{M},\mu \models A(v)$; else $\mathcal{M},\lambda \not\models (\exists v)A(v)$

Exercise 5.35

Let $\mathcal{M} = \langle \mathcal{D}, \mathcal{I}\rangle$ where $\mathcal{D} = \{0, 1, 2, 3\}$ and let λ be an assignment of individual variables and arbitrary names to $\mathcal{D}$. Assume further that $\lambda(x) = 0$. How many x-alternative assignments does λ have? Define them.

Exercise 5.36

Demonstrate for each of the following wffs that it is universally valid. (Hint: demonstrate that it is true at an arbitrary assignment in an arbitrary model.)

(a) Fa ∨ ¬Fa
(b) (x)Fx → Fm
(c) (x)Fx → Fa
(d) (x)Fx → (∃x)Fx
(e) (x)Fx → (∃y)Fy
(f) (x)Fx ↔ ¬(∃x)¬Fx
(g) (x)Fx ∨ (y)Gy → (z)(Fz ∨ Gz)

Exercise 5.37

Demonstrate each of the following entailments.

(a) (x)(Fx → Gx), (x)Fx ⊨ (x)Gx
(b) (x)Fxx ⊨ (x)(∃y)Fxy
(c) (x)¬Fxx, (x)(y)(z)(Fxy → (Fzy → Fxy)) ⊨ (x)(y)(Fxy → ¬Fyx)
(d) (x)(y)(Fxy → ¬Fyx), (x)(y)(z)(Fxy → (Fzy → Fxy)) ⊨ (x)¬Fxx
(e) (x)(Fx → P) ⊣⊨ (∃x)Fx → P
(f) (x)(y)(Fxy → Fyy) ⊨ (x)(y)(Fxy → (∃z)(Fxz ∧ Fzy))

Exercise 5.38

For each of the following claims of entailment, define a quantificational model that falsifies it.

(a) $(x)(\exists y)Fxy \vDash (x)Fxx$
(b) $(x)(y)(Fxy \rightarrow (\exists z)(Fxz \wedge Fzy)) \vDash (x)(y)(Fxy \rightarrow Fyy)$
(c) $(x)Fx \rightarrow (x)Gx \vDash (x)(Fx \rightarrow Gx)$

5.11 The Soundness of L_q

In demonstrating the soundness of L_q, we adopt the same strategy that we used for the corresponding metatheorem for L. That is, we demonstrate that in any L_q-proof, the sequent corresponding to the last line of the proof is universally valid. As before, we demonstrate this by demonstrating, by mathematical induction, that for *every* line of an L_q-proof, the sequent corresponding to the line is universally valid. In fact, however, our proof amounts to an extension of that earlier proof, since L_q incorporates the nine rules of L, simply adding to them the four introduction and elimination rules for the universal and existential quantifiers. Since the truth-conditions for negations, conditionals, conjunctions and disjunctions in quantificational models are exactly the classical truth-conditions tabulated in their truth-tables, we need not repeat the nine cases of the induction step that we have already considered, those in which line k is justified by one of the propositional rules. We take up the inductive step for the remaining, quantificational rules.

Assume that for every line before line k, the assumptions of the line entail its conclusion (H.I.).

(Case 1) Assume that line k is justified by UE. Then the proof has the following lines:

Γ_i (i) $(v)A(v)$ <rule> etc.

.

.

Γ_k (k) A(t)   i,UE.

Let $\mathcal{M} = \langle \mathcal{D}, \mathcal{I} \rangle$ be an arbitrary model and λ an arbitrary assignment of individual variables and arbitrary names to $\mathcal{D}$. Assume that $\mathcal{M},\lambda \vDash B$, for every B in Γ_k. But $\Gamma_k = \Gamma_i$, and by H.I. $\Gamma_i \vDash (v)A(v)$. Therefore $\mathcal{M},\lambda \vDash (v)A(v)$. Therefore, for every v-alternative assignment, μ, $\mathcal{M},\mu \vDash A(v)$. Let ν be an assignment such that $R_v\lambda\nu$ and $\nu(v) = \lambda(t)$. Then $\mathcal{M},\nu \vDash A(v)$. But λ and ν are identical except in their assignment of v. Therefore $\mathcal{M},\lambda \vDash A(t)$. (To put the matter informally, ν assigns v to some object o, so the property A(v) is true of o. But λ assigns the term t to o. Therefore A(t) is true at λ.) But $\mathcal{M} = \langle \mathcal{D}, \mathcal{I} \rangle$ was an arbitrary model and λ an arbitrary assignment. Therefore $\Gamma_k \vDash A(t)$.

(Case 2)   Assume that line k is justified by EI. Then the proof has the following lines:

Γ_i (i)   A(t)   \<rule\> etc.
                 .

                         .

Γ_k (k) $(\exists v)A(v)$ i,EI.

Let $\mathcal{M} = \langle \mathcal{D}, \mathcal{I} \rangle$ be an arbitrary model and λ an arbitrary assignment of individual variables and arbitrary names to $\mathcal{D}$. Assume that $\mathcal{M},\lambda \vDash B$, for every B in Γ_k. But $\Gamma_k = \Gamma_i$, and by H.I. $\Gamma_i \vDash A(t)$. Therefore $\mathcal{M},\lambda \vDash A(t)$. Let ν be an assignment such that $R_v\lambda\nu$ and $\nu(v) = \lambda(t)$. Then $\mathcal{M},\nu \vDash A(v)$. Therefore $\mathcal{M},\lambda \vDash (\exists v)A(v)$. (To put the matter informally, λ assigns t to some object o, so the property A(v) is true of o. But ν assigns the individual variable v to o. Therefore A(v) is true at ν.) But $\mathcal{M} = \langle \mathcal{D}, \mathcal{I} \rangle$ was an arbitrary model and λ an arbitrary assignment. Therefore $\Gamma_k \vDash (\exists v)A(v)$.

(Case 3)   Assume that line k is justified by UI. Then the proof has the following lines:

Γ_i (i)   A(e)   \<rule\> etc.
                 .

                         .

Γ_k (k) (v)A(v)   i,UI.

Let $\mathcal{M} = \langle \mathcal{D}, \mathcal{I} \rangle$ be an arbitrary model and λ an arbitrary assignment of individual variables and arbitrary names to $\mathcal{D}$. Assume that $\mathcal{M},\lambda \models B$, for every B in Γ_k. But $\Gamma_k = \Gamma_i$, and by H.I. $\Gamma_i \models A(e)$. Therefore $\mathcal{M},\lambda \models A(e)$. Let μ be an arbitrary assignment such that $R_v\lambda\mu$. Assume that $\mathcal{M},\mu \not\models A(v)$. Let v be an assignment such that $R_v\lambda v$ and $v(e) = \mu(v)$. Therefore $\mathcal{M},\mu \not\models A(e)$. However, $R_v\lambda v$, λ and v agree in their assignments of all arbitrary names having occurrences in Γ_k. Therefore, for every B in $\Gamma_k = \Gamma_i$, $\mathcal{M},v \not\models B$. But then $\Gamma_i \not\models A(e)$ contrary to hypothesis. Therefore $\mathcal{M},\mu \models A(v)$. But μ was an arbitrary v-alternative of λ. Therefore $\mathcal{M},\lambda \models (v)A(v)$.

(Case 4) Assume that line k is justified by EE. Then the proof has the following lines:

Γ_i (i) $(\exists v)A(v)$ \<rule\> etc.

.

Γ_g (g)   A(e)   A

.

Γ_j (j)   C   \<Rule\> etc.

.

Γ_k (k)   C   i,j,k,EE.

Let $\mathcal{M} = \langle \mathcal{D}, \mathcal{I} \rangle$ be an arbitrary model and λ an arbitrary assignment of individual variables and arbitrary names to $\mathcal{D}$. Assume that $\mathcal{M},\lambda \models B$, for every B in Γ_k. But $\Gamma_k = \Gamma_i$ together with $\Gamma_j - \Gamma_g$, and by H.I. $\Gamma_i \models (\exists v)A(v)$. Therefore $\mathcal{M},\lambda \models (\exists v)A(v)$. Moreover, by the truth-table for the conditional, $\Gamma_j - \Gamma_g \models A(e) \rightarrow C$. And again, since by a stipulation of EE, C contains no occurence of the arbitrary name e and depends upon no assumption other than (g) having an occurrence of e. It follows that the conditional $A(e) \rightarrow C$ depends upon no assumption in which e occurs. Therefore, by the previously demonstrated case, $\Gamma_j - \Gamma_g \models (v)(A(v) \rightarrow C), \Gamma_j - \Gamma_g \models (v)(A(v) \rightarrow C)$, in the conclusion of which C contains no occurrence of v. Therefore $\mathcal{M},\lambda \models (v)(A(v) \rightarrow C)$. But as we have seen (Exercise 5.38), if C contains no occurrence of v, then $(v)(A(v) \rightarrow C) \models (\exists v)A(v) \rightarrow C$. Therefore $\mathcal{M},\lambda \models (\exists v)A(v) \rightarrow C$. Therefore, by the truth-conditions for the conditional, $\mathcal{M},\lambda \models C$.

We can conclude that in any L_q-proof, if for every line earlier than line k, the assumptions at the line entail the conclusion at the line, then the assumptions at line k entail the conclusion at line k. It follows that for every line of an L_q-proof, the assumptions at the line entail the conclusion at the line. In particular therefore this holds of the last line of any L_q-proof. Therefore for any ensemble Σ of L_q wffs, and any L_q wff, A, if $\Sigma \vdash A$, then $\Sigma \vDash A$. That is, L_q is sound with respect to the class of all quantificational models.

5.12 The Completeness of L_q

5.12.1 Propositional and quantificational models

We have made much of the connection between the universal quantifier and conjunction on the one hand, and between the existential quantifier and disjunction on the other. Yet the models for a quantificational language do not evidently resemble generalized truth-tables. As we saw earlier, for our interpretation of the monadic fragment of L_q, we can essentially use truth-tables. In fact the difference between the two lies in the presence of n-place predicates for $n > 1$. However, now that we have had an account of more general Q-models, we can re-describe propositional models more generally to make the connection more explicit. An advantage of this is that in this new idiom we can present a completeness proof for the propositional system L that will serve to introduce, on a smaller scale, the standard technique by which completeness is proved in quantificational systems.

5.12.2 Propositional models revisited: local models

Truth-tables, like Q-models require both an assignment and an interpretation. However, since its atoms lie in the 0-adic fragment of L_q, truth-tables make no mention of a domain of individuals: 0-place predicates are instantiated by states of affairs, not specifically by individuals. Accordingly, the assignment merely assigns truth-values to propositional atoms. If a wff A has n atoms, then there are 2^n such assignments. Each row of a truth-table of a wff A computes, from one of these assignments, a truth-value for every subwff of A.

Now the assignments recorded in a truth-table of A appear to assign truth-values only to the atoms of A, and indeed these are the only truth-values required to compute the truth-value of A. But it is an easy matter to extend the notion of an assignment to one which assigns values to every atomic wff. Such a *global* assignment would enable us to compute a truth-value for every wff of the propositional language. A more general account of propositional semantics than the one found in Chapter 3 of this text might introduce this more general understanding of an assignment. Our account, by contrast could be called *local*: that is, local to a particular wff, or in the case of sequents, local to a finite set of wffs. We can of course *think* of global assignments in truth-tabular terms, but the truth-table of a global assignment would have a countable infinity of columns and an uncountable infinity of rows. An assignment is local to a finite set Σ of wffs (is Σ-local) iff it assigns truth-values to the atoms of the wffs of Σ, and to no other atoms.

For some purposes it is convenient to interpret a Σ-local assignment λ as a representative of an infinite class of global assignments, namely the class of global assignments that agree with λ on the atoms of the wffs of Σ. On this understanding, every row of a truth-table represents an uncountable class of assignments to *At*: namely the class of assignments indistinguishable from λ in their assignment of values to the atomic constituents of wffs in Σ. We can find a counterpart of v-accessibility here as well, but since there is no individual variable v for assignments λ and μ to differ on, there will be a relation R, but no subscript. In this understanding, λ and μ will be accessible, or alternative assignments iff λ and μ are in the same equivalence class of global assignments. We could easily contrive a wholly otiose idiom in which we could say that for every wff ς in Σ, σ is true at λ iff σ is true at every alternative assignment μ. For a set Σ of wffs, we will define a Σ-*local model*, and understand this as a model which assigns a value from the set $\{0, 1\}$ to every atom of every wff of Σ, and interprets every non-atomic wff of Σ.

Definition of a Σ-local model. Let S be a set of propositional wffs. Then a local model $\mathcal{M}$ for Σ is an ordered pair $< 2, \lambda >$, where 2 is the set $\{0, 1\}$ and λ is a function from the atomic constituents of the wffs of Σ into 2, that is, an assignment assigning every atom of every wff of S either the value 0 or the value 1. We define the satisfaction-relation $\models$ inductively by the following:

For atomic A, $\mathcal{M}, \lambda \models A$ if $\lambda(A) = 1$, else $\mathcal{M}, \lambda \not\models A$.

For non-atomic wffs, we adopt the following clauses:

If A is of the form $\neg B$, then $\mathcal{M}, \lambda \models A$ if $\mathcal{M}, \lambda \not\models B$; else $\mathcal{M}, \lambda \not\models A$.

If A is of the form $B \rightarrow C$, then $\mathcal{M}, \lambda \models A$ if $\mathcal{M}, \lambda \not\models B$ or $\mathcal{M}, \lambda \models C$; else $\mathcal{M}, \lambda \not\models A$.

If A is of the form $B \vee C$, then $\mathcal{M}, \lambda \models A$ if $\mathcal{M}, \lambda \models B$ or $\mathcal{M}, \lambda \models C$; else $\mathcal{M}, \lambda \not\models A$.

If A is of the form $B \wedge C$, then $\mathcal{M}, \lambda \models A$ if $\mathcal{M}, \lambda \models B$ and $\mathcal{M}, \lambda \models C$; else $\mathcal{M}, \lambda \not\models A$.

It will be apparent that what we have called a *global model* is a special case of a local model. A global model is a local model for Φ, the set of all propositional wffs.

5.12.3 Soundness of L with respect to local models

Since our account of local models for finite sets is, in effect, a re-description of truth-tables, the soundness proof of Chapter 3, with trivial modifications, provides a soundness proof for L. Those modifications require only references to satisfaction conditions wherever the original proof made reference to truth-tables. The more dramatic divergence of method becomes evident in the proof of completeness to which we now turn.

5.12.4 Completeness of L with respect to local models: strategy

On our understanding of completeness, a system is complete with respect to an interpretation if and only if, for every ensemble Σ and every wff A, if $\Sigma \vDash A$ on the interpretation, then A is provable from Σ in the system L. To understand the strategy for completeness that we are about to adopt, it is necessary to see that this formulation is equivalent to the following:

> For every ensemble Σ and every wff A, if the set Σ, $\neg A$ cannot be satisfied on the interpretation, then the set $\Sigma, \neg A$ is L-inconsistent.

To see the equivalence recall that $\Sigma \vDash A$ if and only if every assignment of truth-values to atoms that makes all of the wffs of Σ true also makes A true. Alternatively, no assignment of truth-values that makes all of the wffs of Σ true makes A false. That is, no assignment satisfies the set Σ, $\neg A$. Then, notice that if $\Sigma \vdash A$, then the set Σ, $\neg A \vdash A \wedge \neg A$. But again, that formulation is equivalent to the following, which is got by transposition. For every ensemble Σ and every wff A, if the set Σ, $\neg A$ is L-consistent, then Σ, $\neg A$ can be satisfied on the interpretation. Accordingly we can prove completeness of L for an interpretation by demonstrating that: every finite L-consistent set of wffs is satisfiable on the interpretation since if every finite L-consistent set is satisfiable, then so is every finite L-consistent set Σ, $\neg A$.

5.12.5 Proving completeness of L for local models: preliminaries

Our strategy is a recipe. We demonstrate, for any finite L-consistent set Δ, how to define a local model that satisfies Δ. We demonstrate how an L-consistent set Δ can be extended to a set Δ' by reference to which we define a local model $\mathscr{M}_{\Delta'}$ which satisfies Δ. The recipe is as follows:

Let $[\Delta]$ be the subwff closure of Δ. (By this is meant the set that includes Δ, and that contains all of the subwffs of wffs in Δ, all of the subwffs of subwffs of wffs in Δ, and so on.)

List the wffs of $[\Delta]$ in some order or other. (It doesn't matter which.) In that order extend Δ, one wff at a time, adding wffs from $[\Delta]$ according to the following rule. If the set resulting from the addition of a wff B from $[\Delta]$ is L-consistent, then add the wff B; if the set resulting from the addition of a wff is L-inconsistent, then add the negation of that wff, $\neg B$.

Since there are only finitely many wffs in [Δ], we must eventually obtain a set that is L-consistent, and the largest L-consistent set capable of being produced in this way on this ordering of [Δ]. We will call this set Δ′.Notice that Δ′ need not be a subset of [Δ], for it may contain negations of wffs in [Δ]. However, if we extend [Δ] to [Δ]⁺, by adding the negations of wffs in [Δ], then we may say that Δ′ is a maximal L-consistent subset of [Δ]⁺. To say that Δ′ is a *maximal* L-consistent subset of [Δ]⁺, is to say that although Δ′ is itself L-consistent, any set resulting from the addition of new wffs of [Δ]⁺ to Δ′ is L-inconsistent.

We must satisfy ourselves that the definition of Δ′ is consistent. That is, we must demonstrate that if the addition of a wff B yields an L-inconsistent set, then the addition of ¬B does not. We can prove this quite generally:

For any L-consistent set, Γ and any wff B, if Γ,B is L-inconsistent, then Γ,¬B is L-consistent. Assume not. Then for some L-consistent set Γ, contradiction C, and wff B, both Γ,B ⊢ C, and Γ,¬B ⊢ C. But then there is an L-proof of C from Γ,B ∨ ¬B, by ∨E. And therefore a proof of C from Γ, since B ∨ ¬B can be introduced by LEM. That is, contrary to our initial hypothesis, Γ is L-inconsistent. We conclude that Δ′ is L-consistent.

5.12.5.1 δ-Consistency (delta-Consistency)

We note that Δ′ has another consistency property, which we may label δ-*consistency*.

A set Σ of wffs is δ-consistent iff for every wff A and every wff B, if A ∨ B is in Σ, then either A is in Σ or B is in Σ. To see that Δ′ is δ-consistent, suppose that A ∨ B is in Δ′, but that neither A nor B is in Δ′. Then A ∨ B is in [Δ], and therefore so are both A and B. But then, by the definition of Δ′, both ¬A and ¬B are in Δ′. But A ∨ B and ¬A yield a proof of B (by MTP). Thus Δ′ ⊢ B and Δ′ ⊢ ¬B. Therefore Δ′ ⊢ B ∧ ¬B. Therefore Δ′ is L-inconsistent, contrary to its definition.

5.12.6 Defining $\mathcal{M}_{\Delta'}$ and putting it to work

$\mathcal{M}_{\Delta'}$ is the local model $<2, \lambda_{\Delta'}>$ where $\lambda_{\Delta'}$ is defined by:

For every atom P_i in $[\Delta]$, $\lambda_{\Delta'}(P_i) = 1$ if P_i is in Δ'; $\lambda_{\Delta'}(P_i) = 0$ else.

We must now demonstrate the following:

For every A in $[\Delta]$, $\mathcal{M}_{\Delta'}, \lambda_{\Delta'} \vDash A$ if and only A is in Δ'.

We demonstrate the result by an induction on the length of A, which, as usual, we understand as the number of symbols in A.

Basis: A is P_i for some i.

Then $\mathcal{M}_{\Delta'}, \lambda_{\Delta'} \vDash A$ iff $\lambda_{\Delta'}(P_i) = 1$ (by definition of $\vDash$). But $\lambda_{\Delta'}(P_i) = 1$ iff P_i is in Δ'. Therefore if $A = P_i$, then $\mathcal{M}_{\Delta'}, \lambda_{\Delta'} \vDash A$ if and only A is in Δ'.

For the hypothesis of induction (H.I.), assume the biconditional for all wffs in $[\Delta]$ of length less than k. Assume A, in $[\Delta]$, is of length k. Then A is of one of four forms: $\neg B$, $B \rightarrow C$, $B \wedge C$, or $B \vee C$. Consider these cases in turn.

Case 1. A is of the form $\neg B$.

Assume that $\neg B$ is in Δ'; then B is not in Δ'. By H.I., $\mathcal{M}_{\Delta'}, \lambda_{\Delta'} \nvDash B$. Therefore $\mathcal{M}_{\Delta'}, \lambda_{\Delta'} \vDash \neg B$.
Assume $\neg B$ is not in Δ'. Then B is in Δ'. By H.I., $\mathcal{M}_{\Delta'}, \lambda_{\Delta'} \vDash B$. Therefore $\mathcal{M}_{\Delta'}, \lambda_{\Delta'} \nvDash \neg B$.

Case 2. A is of the form $B \rightarrow C$.

Assume that $B \rightarrow C$ is in Δ'; then either B is not in Δ' or C is. By H.I., $\mathcal{M}_{\Delta'}, \lambda_{\Delta'} \nvDash B$ or $\mathcal{M}_{\Delta'}, \lambda_{\Delta'} \vDash C$. Therefore $\mathcal{M}_{\Delta'}, \lambda_{\Delta'} \vDash B \rightarrow C$.
Assume $B \rightarrow C$ is not in Δ'. Then B is in Δ' and C is not. By H.I., $\mathcal{M}_{\Delta'}, \lambda_{\Delta'} \vDash B$ and $\mathcal{M}_{\Delta'}, \lambda_{\Delta'} \nvDash C$. Therefore $\mathcal{M}_{\Delta'}, \lambda_{\Delta'} \nvDash B \rightarrow C$.

Case 3. A is of the form B $\wedge$ C.

Assume that B $\wedge$ C is in Δ'; then both B and C are in Δ'. By H.I., $\mathcal{M}_{\Delta'}, \lambda_{\Delta'} \vDash$ B and $\mathcal{M}_{\Delta'}, \lambda_{\Delta'} \vDash$ C. Therefore $\mathcal{M}_{\Delta'}, \lambda_{\Delta'} \vDash$ B $\wedge$ C.
Assume B $\wedge$ C is not in Δ'. Then either B is not in Δ' or C is not. By H.I., $\mathcal{M}_{\Delta'}, \lambda_{\Delta'} \nvDash$ B or $\mathcal{M}_{\Delta'}, \lambda_{\Delta'} \nvDash$ C. Therefore $\mathcal{M}_{\Delta'}, \lambda_{\Delta'} \nvDash$ B $\wedge$ C.

Case 4. A is of the form B $\vee$ C.

Assume B $\vee$ C is in Δ'. Then either B is in Δ' or C is. By H.I., $\mathcal{M}_{\Delta'}, \lambda_{\Delta'} \vDash$ B or $\mathcal{M}_{\Delta'}, \lambda_{\Delta'} \vDash$ C. Therefore $\mathcal{M}_{\Delta'}, \lambda_{\Delta'} \vDash$ B $\vee$ C.
Assume that B $\vee$ C is not in Δ'; then neither B nor C is in Δ'. By H.I., $\mathcal{M}_{\Delta'}, \lambda_{\Delta'} \nvDash$ B and $\mathcal{M}_{\Delta'}, \lambda_{\Delta'} \nvDash$ C. Therefore $\mathcal{M}_{\Delta'}, \lambda_{\Delta'} \nvDash$ B $\vee$ C.

This completes the induction. For every A in [Δ], $\mathcal{M}_{\Delta'}, \lambda_{\Delta'} \vDash$ A if and only A is in Δ'. But every wff of Δ is a member of [Δ]. Therefore, for every A in Δ, $\mathcal{M}_{\Delta'}, \lambda_{\Delta'} \vDash$ A. That is, $\mathcal{M}_{\Delta'}, \lambda_{\Delta'}$ satisfies Δ. But Δ was an arbitrary finite L-consistent set of wffs. We may therefore conclude that every finite L-consistent set of wffs is satisfiable in a local model. Therefore, L is complete.

5.12.7 The strategy modified for the quantificational setting

Evidently, such a strategy must be adapted to the language of any system S to which it is applied. After all, what we wish to demonstrate is that every finite *S-consistent* set of wffs is satisfiable *in an S-model*. So we must ask ourselves: what are the relevant differences to be taken into account in the adaptation?

Recall first that we introduced the universal quantifier as a generalized conjunction and the existential quantifier as a generalized disjunction: *generalized* because if we are to think of them as conjunction and disjunction respectively, then we must countenance infinite conjunctions and disjunctions. Again recall that when, in the preceding completeness proof for L, we expanded Δ to a δ-consistent set Δ', we were able to demonstrate that for any wffs A, B, if A $\vee$ B is in Δ', then one or the other of A, B is also in Δ'. We called this property δ-consistency. In that demonstration, δ-consistency underwrote our truth-definition: we could not have A $\vee$ B true in the model, without having one or the other of A, B true in the model. δ-consistency guaranteed that that would be so.

5.12.7.1 ω-Consistency

The existential quantifier imposes a corresponding requirement. As in the propositional case, we will demonstrate that any finite consistent set can be satisfied in a quantificational model. For a particular finite consistent set Σ, we extend it to a maximal L_q-consistent subset, Σ', of Φ, (though in this case, not *just* any maximal consistent subset of Φ will do the trick). As in the propositional case we will want a wff to be true in the model we define if and only if the wff is a member of Σ'. Now suppose that a wff of the form $(\exists v)A(v)$ is a member of the set Σ'. Then we will want there to be some individual, i, in the domain of the model, of which A is true. That is, we will want the individual variable v to have been assignable to some individual that will make $(\exists v)A(v)$ true. But again, truth will be defined by membership in a set of wffs. Therefore we want there to be some term, t such that the wff A(t) is in Σ'. The individual to which t is assigned is then said to be a *witness* to the truth of $(\exists v)A(v)$, and A(t) a witness wff. A consistent set that has this property, namely that for every wff of the form $(\exists v)A(v)$ in the set, there is a wff of the form A(t) in the set, is said to be ω-consistent (omega-consistent). However, unlike the propositional case in which any maximal L-consistent subset of Φ can be demonstrated to be δ-consistent, there is no such guarantee in the quantificational case that a maximal L_q-consistent set Σ' will be ω-consistent. If we want the set to provide for witnesses to the truth of its existentially quantified wffs, then we must define the set in such a way that those witnesses will be available. For every wff of the form $(\exists v)A(v)$ in the set, we must ensure that a suitable wff of the form A(t) is also in the set. We state the definition of ω-consistency for future reference:

A set Σ of wffs is ω-consistent iff Σ is consistent, and for every wff of the form $(\exists v)A(v)$ in Σ, there is a wff of the form A(t) in Σ.

5.12.7.2 The domain

Recall that the atoms of Φ_q are analysable into a predicate symbol P, and a succession $t_1 \ldots t_n$, of terms. Accordingly, if a quantificational model is to give truth-values to atomic wffs, it requires a domain of individuals that will provide an interpretation of predicate symbols and proper names, and allow assignments of individual variables and arbitrary names.

Any non-empty set of objects is eligible to be the domain of a Q-model. So, for example, the set of individual variables would be eligible; so would the set of all arbitrary names; so would the set of all proper names, for individual variables and names are themselves objects. For this model, we will take as our domain the set of all terms. That is, every object in the domain of the model will be either a proper name or an arbitrary name, and every proper name and every arbitrary name will be in the domain. In the model we define, we will be able to assign individual variables and arbitrary names to objects in this domain. *Any such assignment would be acceptable*, and we must bear that in mind as we proceed, but for the purposes of our demonstration, *we will interpret every proper name as itself, and assign every arbitrary name to itself.* Each individual variable v_i will be assigned to the arbitrary name e_i.

Recall that the goal of this construction is that for any finite L_q-consistent set of wffs Σ, we should be able to define a model in which Σ is satisfied at some assignment. Eventually, given the model that we define, we must also define an assignment λ at which all of the wffs of Σ are true in the model. We will want to be able to demonstrate this by an induction. Accordingly we will want to define the interpretation and the assignment in such a way as to make all of the atoms of Σ true in some trivial and obvious way. The definitions of domain and assignments that we have been contemplating will contribute to that goal. The role of the definition of the interpretation will have to await two other components of our demonstration.

5.12.8 Maximal L_q-consistent subsets of Φ_q

Recall a remark in our discussion of Σ-local models, to the effect that for a finite Σ, we can think of a Σ-local assignment as a representative of an infinite class of global assignments. The reason is that for every atom having no occurrence in Σ, we can distinguish two classes of assignments: those that assign the atom to 1, and those that assign it to 0. If there are infinitely many such 'fresh' atoms, the number of such classes will be the value of 2 raised to an infinite power. Evidently, if Σ is Φ_q itself, then a Σ-local assignment will be unique: there will be no atoms left over by which to distinguish classes of assignments. A global assignment can be defined by reference to a maximal consistent subset of Φ_q. We will therefore demonstrate that an L_q-consistent set can be extended to a maximal L_q-consistent subset of Φ_q. The demonstration requires that Φ_q can be enumerated.

5.12.8.1 Enumerating Φ_q

No new techniques are required to demonstrate that Φ_q can be enumerated. We merely adapt the method introduced in Chapter 3 to the requirements of a quantificational language. We define a translation that assigns to the symbols "(", ")", "$\to$", "$\neg$", "$\vee$", "$\wedge$", "$\exists$" the first ten powers of 10 ($10^0 - 10^9$), then we order the *sets* of predicate symbols, of proper names, of arbitrary names, and of individual variables in some determinate way. Finally we assign the members of those sets to the higher powers of 10, cycling through the sets in the adopted order. Thus the first entries of the sets will be assigned the powers $10^{10} - 10^{13}$; the second, $10^{14} - 10^{17}$, and so on. We then arithmetize each wff of Φ_q as the concatenation of the arithmetic translations of its component symbols. The enumeration of Φ_q will be the natural ordering of the arithmetizations of its members (that is, the order in which the numbers they name occur among the natural (or counting) numbers.)

5.12.8.2 The first component: the Extension Lemma

Assume that Γ is an L_q-consistent set of wffs. We must demonstrate that Γ can be extended to a maximal L_q-consistent subset of Φ_q, that is, that there exists a maximal L_q-consistent subset of Φ_q that has Γ as a subset. We do this by defining a sequence of sets $\Gamma, \Gamma_1 \ldots \Gamma_i \ldots$, then demonstrating that the union of that sequence of sets (that is, the set of all wffs that are in at least one of the sets) is a maximal L_q-consistent set of wffs. Each successive set adds a wff from the enumeration of Φ_q if it can consistently do so, else it omits it. The sequence of sets is defined as follows:

$\Gamma_0 = \Gamma$
$\Gamma_1 = \Gamma_0, A_1$ if that set is L_q-consistent;
      else, $\Gamma_1 = \Gamma_0$.

In general, for every counting number i,

$\Gamma_i = \Gamma_{i-1}, A_i$ if that set is L_q-consistent;
      else, $\Gamma_i = \Gamma_{i-1}$.

We define Γ^+ as the union of all of the sets in the sequence.

Now, we have merely *defined* Γ^+. There remains to check that Γ^+ is a maximal L_q-consistent set of wffs. This requires us to demonstrate

(a) that Γ^+ is L_q-consistent, and

(b) that any set properly extending Γ^+, (that is, any set that adds a wff not already in Γ^+), is L_q-inconsistent.
We perform these checks in order.

(a) Assume that Γ^+ is L_q-inconsistent. Then $\Gamma^+ \vdash_{L_q} P \wedge \neg P$. But then, by the finiteness property of $\vdash_{L_q}$, there is some finite subset Γ^- of Γ^+ such that $\Gamma^- \vdash_{L_q} P \wedge \neg P$. Every wff of Γ^- has a place in the enumeration of Φ_q. Let the wff of Γ^- that occurs later in that enumeration of Φ_q than any other wff of Γ^- be the wff A_j. Then the set Γ_{j-1}, A_j is L_q-inconsistent. Therefore, by the definition of the sequence, $\Gamma_j = \Gamma_{j-1}$. It follows that A_j is not in Γ^+, and therefore that Γ^- is not a subset of Γ^+, contrary to hypothesis.

(b) Assume that some wff B is not a member of Γ^+. Let B be A_k in the enumeration. Then, by the definition of the sequence of sets Γ_i, the set Γ_{k-1}, A_k is L_q-inconsistent. But Γ_{k-1}, A_k is a subset of Γ^+, B. Therefore (by monotonicity,) Γ^+, B is L_q-inconsistent.

Since Γ was an arbitrary L_q-consistent set of wffs, we may conclude that every L_q-consistent set of wffs can be extended to a maximal L_q-consistent set of wffs.

5.12.9 Properties of maximal L_q-consistent sets of wffs

The uses to which we will want to put maximal L_q-consistent sets of wffs depend upon certain demonstrable properties of such sets. We demonstrate them here:

5.12.9.1 Deductive closure

Maximal L_q-consistent sets of wffs are closed under $\vdash_{L_q}$. To see this, let Δ be a maximal L_q-consistent set. Assume that $\Delta \vdash_{L_q} A$, but that A is not in Δ. Then $\Delta \vdash_{L_q} \neg A$. But then $\Delta \vdash_{L_q} A \wedge \neg A$, contrary to assumption.

5.12.9.2 Every L_q-theorem is in every L_q-maximal consistent set

Since every L_q-theorem is provable from the empty set, by monotonicity, every L_q-theorem is provable from every L_q-maximal consistent set. By deductive closure, every L_q-theorem is in every L_q-maximal consistent set.

5.12.9.3 For any L_q-maximal consistent set Δ and for any wff A, either A is in Δ, or $\neg$A is in Δ

To see this, let Δ be a maximal L_q-consistent set and A a wff. Assume that $\neg$A is not in Δ. Then $\Delta, \neg A \vdash_{L_q} P \wedge -P$. Therefore $\Delta \vdash_{L_q} A$. Therefore A is in Δ, by deductive closure.

5.12.9.4   For any L$_q$-maximal consistent set Δ and for any wffs A and B, A → B is in Δ if and only if either A is not in Δ, or B is in Δ

Let Δ be an L$_q$-maximal consistent set. Assume that ¬A is in Δ. Then by deductive closure, A → B is in Δ. Assume that B is in Δ. Then by deductive closure, A → B is in Δ. Now assume that A → B is in Δ, but that ¬A is not in Δ. Then A is in Δ, by the previous property. But then Δ ⊢$_{L_q}$ B. Therefore B is in Δ.

5.12.10   ω-Consistency again

For our purposes, it is insufficient to demonstrate that every L$_q$-consistent set of wffs has a maximal L$_q$-consistent extension. Our desired demonstration requires that every finite L$_q$-consistent set of wffs has a maximal L$_q$-ω-consistent extension. This requires a further demonstration, namely that every finite L$_q$-consistent set Δ of wffs has an extension Δ′ that has a maximal L$_q$-ω-consistent extension. To demonstrate that this is so, we assume a finite L$_q$-consistent set Δ of wffs, and then define a consistent extension, Δ′ of Δ such that any maximal L$_q$-consistent extension Δ⁺ of Δ′ will be ω-consistent. Because we will have prepared Δ′ in such a way that every maximal L$_q$-consistent extension Δ⁺ of Δ′ will, as it were, *automatically*, be ω-consistent, we will refer to Δ′ as *ω-primed*. Recall that an L$_q$-ω-consistent set is an L$_q$-consistent set that contains a *witness wff*, A(e), e being an arbitrary name, for every wff of the form (∃v)A(v). Our strategy will be to demonstrate that we can extend our finite L$_q$-consistent set Δ of wffs to an ω-primed set Δ′ containing a sufficient set of conditionals of the form (∃v)A(v) → A(e), that we can depend upon the deductive closure of maximal consistent extensions of Δ′ to provide a witness wff A(e) for every wff of the form (∃v)A(v) that is a member of that maximal consistent extension.

Once again we will define a series of sets that extend Δ to a consistent ω-primed set Δ′. To define the series, we will have recourse to our original enumeration of Φ$_q$. Since every wff of the form (∃v)A(v) occupies a place in that enumeration, that enumeration induces an enumeration of the wffs of that form: namely the enumeration that preserves the order of their appearance in the original enumeration. For convenience, we refer to that enumeration of wffs of the form $(\exists v)A(v)$ as *the E-enumeration*

$$E_1, \ldots, E_i, \ldots$$

Now for each wff, E_i in that E-enumeration, there will be a set of witness wffs

$$W_{i_1}, \ldots, W_{i_j}, \ldots$$

of the related form $A(e)$. We call the arbitrary name e, *the replacement term*. Again, for each wff in the E-enumeration, we enumerate its witness wffs using the enumeration induced by our original enumeration of F_q. Thus with each wff beginning with an existential quantifier in the E-enumeration, there is associated an enumerated set of all of its witness wffs. As an example, the wff $(\exists x)Fx$ will appear somewhere in the E-enumeration, and among the enumerated witness wffs for $(\exists x)Fx$ will appear the available witness wffs, Fa, Fa′, and so on.

We are now in a position to define the series of sets that will yield an ω-primed extension for any finite L_q-consistent set Δ.

$\Delta_0 = \Delta$

$\Delta_1 = \Delta, E_1 \rightarrow W_{1_i}$ where W_{1_i} is the first witness wff in its enumeration whose replacement term has no occurrences in Δ_0.

.

.

.

$\Delta_k = \Delta_{k-1}, E_k \rightarrow W_{k_j}$ where W_{k_j} is the first witness wff in its enumeration whose replacement term has no occurrences in Δ_{k-1}.

.

.

.

Δ' is the union of that series of sets.

5.12.10.1 The L_q-consistency of Δ'

Evidently Δ' is ω-primed. To demonstrate that Δ' is also L_q-consistent, it is sufficient to show that if a set in the series is consistent, then the next set in the series is also consistent, in other words, that the addition of $E_k \to W_{k_j}$ preserves consistency if W_{k_j} is a "fresh" arbitrary name. The demonstration is as follows:

Assume that Δ_{k+1} is L_q-inconsistent.
Let the conditional that Δ_{k+1} adds be $(\exists v)A(v) \to A(e)$.
Since Δ_k is finite, we can form its conjunction. Let K be that conjunction.
Then the following wff is a theorem of L_q.

$$K \to \neg((\exists v)A(v) \to A(e))$$

However, since there are no occurrences of the arbitrary name e in K, it follows that the wff

$$K \to (v')\neg((\exists v)A(v) \to A(v'))$$

(where v' has no occurrences in $(\exists v)A(v) \to A(e)$) is also a theorem of L_q. Hence, so is the wff

$$K \to \neg(\exists v')((\exists v)A(v) \to A(v'))$$

and so is the wff

$$(\exists v')((\exists v)A(v) \to A(v')) \to \neg K.$$

But

$$(\exists v')((\exists v)A(v) \to A(v'))$$

is a theorem of L_q. Therefore, so is

$$\neg K.$$

It follows that Δ_k is L_q-inconsistent. Therefore the addition of $E_k \to W_{k_j}$ to Δ_{k-1} preserves L_q-consistency if W_{k_j} is a "fresh" arbitrary name.
Since by assumption Δ is L_q-consistent, we can infer that Δ' is L_q-consistent. Since Δ was an arbitrary finite L_q-consistent set, we may infer

that every finite L_q-consistent set of wffs has an L_q-consistent, ω-primed extension.

Exercise 5.39

Document the foregoing demonstration, by reference to previously proved sequents.

5.12.11 Constructing the model $\mathscr{M}_\Sigma$

It follows from what we have demonstrated so far that

Every finite L_q-consistent set Σ of wffs has a maximal L_q-ω-consistent extension.

The reasoning is as follows.

Every finite L_q-consistent set Σ of wffs has an L_q-consistent, ω-primed extension Σ'. Therefore every finite L_q-consistent set Σ of wffs has a maximal L_q-ω-primed extension, Σ^+. Now suppose that $(\exists v)A(v)$ is in Σ^+. But, for some arbitrary name e, $(\exists v)A(v) \rightarrow A(e)$ is in Σ^+. Σ^+ is deductively closed. Therefore $A(e)$ is in Σ^+.

It follows that, if our goal is to demonstrate that every L_q-consistent set Σ has a model, it will be sufficient to demonstrate that

Every maximal L_q-ω-consistent set has a model.

That is how we proceed. Given a maximal L_q-ω-consistent set, Σ, we define a model, $\mathscr{M}_\Sigma$, that satisfies Σ. Let Σ be a maximal L_q-ω-consistent set. We define a model

$$\mathscr{M}_\Sigma = <\mathscr{D}_\Sigma, \mathscr{I}_\Sigma>$$

as follows:

$\mathscr{D}_\Sigma$ (the Σ-domain) is the set of terms.

$\mathscr{I}_\Sigma$ (the Σ-interpretation)

interprets every proper name m_i as itself.

interprets every n-ary predicate symbol P to the set of n-tuples $\langle t_1, \ldots, t_n \rangle$ such that $Pt_1 \ldots t_n$ is in Σ.

On this model, we define an assignment:

λ_Σ (the Σ-assignment) assigns each arbitrary name to itself and each individual variable v_i to the arbitrary name e_i.

5.12.12 The main result

We demonstrate that for every wff A in Φ_{L_q},

$$\mathscr{M}_\Sigma, \lambda_\Sigma \vDash A \text{ iff } A \text{ is in } \Sigma.$$

The demonstration is by induction on the degree of complexity of A (that is, on the depth of nestings of connectives in A.)

Basis

A is atomic; that is, A is of the form $Pt_1 \ldots t_m$ for some m. By the truth-condition for atoms, $\mathscr{M}_\Sigma, \lambda_\Sigma \vDash Pt_1 \ldots t_m$ iff $\langle \lambda_\Sigma(t_1) \ldots \lambda_\Sigma(t_m) \rangle$ is in $\mathscr{T}_\Sigma(P)$. But for each n, $\mathscr{T}_{n_\Sigma}(P)$ is the set of n-tuples, $\langle t_1 \ldots t_n \rangle$ such that $Pt_1 \ldots t_n$ is in Σ. Therefore if A is atomic, then $\mathscr{M}_\Sigma, \lambda_\Sigma \vDash A$ iff A is in Σ.

Inductive Step

Assume that the biconditional holds for all wffs of degree of complexity less than k. Assume that A is of complexity degree k. Then A is of the form $\neg B$, or $B \to C$, or $B \vee C$, or $B \wedge C$, or $(v)B(v)$, or $(\exists v)B(v)$. We leave the four propositional cases as

Exercise 5.40

Demonstrate that if A is of the form $\neg B$, or $B \to C$, or $B \vee C$, or $B \wedge C$, then $\mathscr{M}_\Sigma, \lambda_\Sigma \vDash A$ iff A is in Σ.

We consider only the two quantificational cases.

Assume that A is of the form (v)B(v).

Assume that (v)B(v) is in Σ. We must demonstrate that $\mathscr{M}_\Sigma, \lambda_\Sigma \vDash$ (v)B(v). That is, we must demonstrate that for every v-alternative assignment μ, $\mathscr{M}_\Sigma, \mu \vDash$ B(v). Consider an arbitrary v-alternative assignment μ. Assume that μ assigns v to t. But $\vdash_{L_q}$ (v)B(v) → B(t). Therefore (v)B(v) → B(t) is in Σ. Therefore, B(t) is in Σ. Moreover $\lambda_\Sigma(t) =$ t. By the hypothesis of induction, $\mathscr{M}_\Sigma, \lambda_\Sigma \vDash$ B(t). Therefore $\mathscr{M}_\Sigma, \mu \vDash$ B(v). Therefore $\mathscr{M}_\Sigma, \lambda_\Sigma \vDash$ (v)B(v).

Assume that (v)B(v) is not in Σ. We must demonstrate that $\mathscr{M}_\Sigma, \lambda_\Sigma \nvDash$ (v)B(v). That is, we must demonstrate that for some v-alternative assignment μ, $\mathscr{M}_\Sigma, \mu \nvDash$ B(v). Since Σ is a maximal L_q-consistent set, we may infer that ¬(v)B(v) is in Σ, and therefore (∃v)¬B(v) is in Σ. But Σ is ω-consistent. Therefore there is some arbitrary name e such that ¬B(e) is in Σ. Then $\mathscr{M}_\Sigma, \lambda_\Sigma \vDash$ ¬B(e). Let μ be the v-alternative assignment that assigns v to e. Then $\mathscr{M}_\Sigma, \mu \vDash$ ¬B(x). Therefore, $\mathscr{M}_\Sigma, \lambda_\Sigma \nvDash$ (v)B(v).

Assume that A is of the form (∃v)B(v).

Assume that (∃v)B(v) is in Σ. We must demonstrate that $\mathscr{M}_\Sigma, \lambda_\Sigma \vDash$ (∃v)B(v). That is, we must demonstrate that for some v-alternative assignment μ, $\mathscr{M}_\Sigma, \mu \vDash$ B(v). Σ is ω-consistent. Therefore there is some arbitrary name e such that B(e) is in Σ. Then $\mathscr{M}_\Sigma, \lambda_\Sigma \vDash$ B(e). Let μ be the v-alternative assignment that assigns v to e. Then $\mathscr{M}_\Sigma, \mu \vDash$ B(v). Therefore, $\mathscr{M}_\Sigma, \lambda_\Sigma \vDash$ (∃v)B(v).

Assume that (∃v)B(v) is not in Σ. We must demonstrate that $\mathscr{M}_\Sigma, \lambda_\Sigma \nvDash$ (∃v)B(v). That is, we must demonstrate that for every v-alternative assignment μ, $\mathscr{M}_\Sigma, \mu \nvDash$ B(v). Since Σ is a maximal L_q-consistent set, we may infer that (v)¬B(v) is in Σ. Consider an arbitrary v-alternative assignment μ. Assume that μ assigns v to t. But $\vdash_{L_q}$ (v)¬B(v) → ¬B(t). Therefore, ¬B(t) is in Σ. Moreover $\lambda_\Sigma(t) =$ t. By the hypothesis of induction, $\mathscr{M}_\Sigma, \lambda_\Sigma \vDash$ ¬B(t). Therefore $\mathscr{M}_\Sigma, \mu \vDash$ ¬B(v). Therefore $\mathscr{M}_\Sigma, \mu \nvDash$ B(v). Therefore, $\mathscr{M}_\Sigma, \lambda_\Sigma \nvDash$ (∃v)B(v).

This completes the induction. We conclude that for every wff A in Φ_{L_q}, $\mathcal{M}_\Sigma$, $\lambda_\Sigma \vDash A$ iff A is in Σ. Therefore $\mathcal{M}_\Sigma$ satisfies Σ. But Σ was an arbitrary maximal L_q-ω-consistent set. Therefore every maximal L_q-ω-consistent set is satisfiable in a quantificational model. But every finite L_q-consistent set can be extended to a maximal L_q-ω-consistent set. Therefore every finite L_q-consistent set is satisfiable in a quantificational model. Therefore L_q is complete with respect to the class of all quantificational models.

A

Normal Forms

A.1 Substitution of Provable Equivalents

If A ⊣⊢ B, then all of the following equivalences hold in L.

(a)   ¬A ⊣⊢ ¬B
(b)   A → C ⊣⊢ B → C
(c)   C → A ⊣⊢ C → B
(d)   A ∨ C ⊣⊢ B ∨ C
(e)   C ∨ A ⊣⊢ C ∨ B
(f)   A ∧ C ⊣⊢ B ∧ C
(g)   C ∧ A ⊣⊢ C ∧ B

Consider them in turn.

(a)   Assume that A ⊣⊢ B. Then A ⊢ B. Then the proof of ¬A from ¬B is the assumption of ¬B followed by the proof of B from A, followed by the entry A → B, justified by CP, followed by the entry ¬A, justified by MTT. Likewise B ⊢ A, and the proof of ¬B from ¬A is the assumption of ¬A, followed by the proof of A from B, followed by the entry B → A, justified by CP, followed by the entry ¬B, justified by MTT.

(b)   Assume that A ⊣⊢ B. Then A ⊢ B. Then the proof of A → C from B → C is the assumption of B → C, followed by the proof of B from A, followed by the entry C, justified by MPP, followed by the entry A → C, justified by CP. Likewise, B ⊢ A. Then the proof of

B → C from A → C is the assumption of A → C, followed by the proof of A from B, followed by the entry C, justified by MPP, followed by the entry B → C, justified by CP.

(c)   Assume that A ⊣⊢ B. Then A ⊢ B. Then the proof of C → B from C → A is the assumption of C → A, followed by the assumption of C followed by the proof of B from A, followed by the entry C → B, justified by CP. Likewise, B ⊢ A. Then the proof of C → A from C → B is the assumption of C → B, followed by the assumption of C followed by the proof of A from B, followed by the entry C → A, justified by CP.

(d)   Assume that A ⊣⊢ B. Then A ⊢ B. Then the proof of B ∨ C from A ∨ C is the assumption of A ∨ C, followed by the proof of B from A, followed by the entry B ∨ C, justified by ∨I, followed by the assumption of C followed by the entry B ∨ C, justified by ∨I, followed by the entry B ∨ C, justified by ∨E. Likewise, B ⊢ A. Then the proof of A ∨ C from B ∨ C is the assumption of B ∨ C, followed by the proof of A from B, followed by the entry A ∨ C, justified by ∨I, followed by the assumption of C followed by the entry A ∨ C, justified by ∨I, followed by the entry A ∨ C, justified by ∨E.

(e)   The demonstration requires only a trivial amendment of (d).

(f)   Assume that A ⊣⊢ B. Then A ⊢ B. Then the proof of B ∧ C from A ∧ C is the assumption of A ∧ C, followed by the proof of B from A (justified by ∧E), followed by the entry C, justified by ∧E, followed by the entry B ∧ C, justified by ∧I. Likewise, B ⊢ A. Then the proof of A ∧ C from B ∧ C is the assumption of B ∧ C, followed by the proof of A from B (justified by ∧E), followed by the entry C, justified by ∧E, followed by the entry A ∧ C, justified by ∧I.

(g)   The demonstration requires only a trivial amendment of (f).

It follows from these demonstrations that if, in any wff, we substitute for some well-formed string, another well-formed string that is provably equivalent to it, then the wff resulting from the substitution is provably equivalent to the original wff. The point may not be immediately apparent. However, consider as an example, the wff

$$A = (P \rightarrow (Q \rightarrow R)) \rightarrow ((P \rightarrow Q) \rightarrow (P \rightarrow R)).$$

The shortest well-formed strings of A are its propositional letters P, Q, and R. Now the only wff provably equivalent to an atom is the atom itself, so the only relevant substitution for those strings is the vacuous one which produces exactly A, which is provably equivalent to A by the Rule of Assumption. Now consider the next shortest well-formed strings of A. These are the conditionals:

1. $(Q \rightarrow R)$

2. $(P \rightarrow Q)$

3. $(P \rightarrow R)$

Consider them in turn.

(1) Suppose that we substitute for $(Q \rightarrow R)$ some provably equivalent wff, say $(\neg Q \lor R)$. Then by demonstration (c) above, we may infer that $(P \rightarrow (Q \rightarrow R))$ is provably equivalent to $(P \rightarrow (\neg Q \lor R))$. So the contemplated substitution of $(\neg Q \lor R)$ for $(Q \rightarrow R)$ replaces $(P \rightarrow (Q \rightarrow R))$ by a provably equivalent wff. But $(P \rightarrow (Q \rightarrow R))$ is the antecedent of A. Therefore, by the demonstration (b) above, A is provably equivalent to the wff that results from the proposed substitution of $(\neg Q \lor R)$ for $(Q \rightarrow R)$.

(2) Suppose that we substitute $(\neg P \lor Q)$ for $(P \rightarrow Q)$. Then by demonstration (b) above, $((P \rightarrow Q) \rightarrow (P \rightarrow R))$ is provably equivalent to $((\neg P \lor Q) \rightarrow (P \rightarrow R))$. By demonstration (c) A is provably equivalent to $(P \rightarrow (Q \rightarrow R)) \rightarrow ((\neg P \lor Q) \rightarrow (P \rightarrow R))$.

(3) Suppose that we substitute $(\neg P \lor R)$ for $(P \rightarrow R)$. Then by demonstration (c), $((P \rightarrow Q) \rightarrow (P \rightarrow R))$ is provably equivalent to $((P \rightarrow Q) \rightarrow (\neg P \lor R))$. By demonstration (b) A is provably equivalent to $(P \rightarrow (Q \rightarrow R)) \rightarrow ((P \rightarrow Q) \rightarrow (\neg P \lor R))$.

The next shortest well-formed string is (P → (Q → R)), but, by (b), any substitution of a provably equivalent wff for that string will yield a wff provably equivalent to A

Then consider the well-formed string ((P → Q) → (P → R)). By (c), any substitution of a provably equivalent wff for that string will yield a wff provably equivalent to A.

Finally, we must consider A itself. Evidently this case is trivial. The substitution for A of a wff provably equivalent to A yields a wff provably equivalent to A.

Now this has been an illustration, not a demonstration. But it is generalizable to all wffs. The well-formed strings from which a wff is composed form a branching structure, with the wff itself as the uppermost node, and the occurrences of its propositional letters as the nodes at the bottom. Since the language has only the connectives →, ¬, ∧ and ∨, every branching is to wffs that fall into the categories of (a) to (g) above. Any substitution of a provable equivalent for any well-formed string within the wff yields a substitution of a provably equivalent wff at every superior node in that branching structure of well-formed strings.

Exercise A.41

Demonstrate, by citing a counterexample, the falsity of the following claim:

Any substitution of a provable equivalent for any well-formed string within the wff yields a substitution of a provably equivalent wff at every *inferior* node in that branching structure of well-formed strings.

A consequence of the foregoing is summed up in the *Rule for the Substitution of Provable Equivalents (SPE)*:

Let A be a wff and B a well-formed string of A. Let B′ be a wff such that B ⊣⊢ B′, and A′ be the wff resulting from the substitution of B′ for an occurrence of B in A. Then A ⊣⊢ A′.

Note the difference between SPE and uniform substitution. Uniform substitution permits the substitution of *any* wff at all for *every* occurrence of any *propositional letter* in A and preserves theoremhood. Uniform substitution permits only substitutions for propositional letters, permits any wff to be substituted, but requires that the substitution be made for every occurrence of the propositional letter. By contrast, SPE permits substitution for a single occurrence of an arbitrary well-formed string. It permits no non-vacuous substitution for propositional letters, does permit substitution for arbitrary well-formed strings, does not require that the substitution be made for every occurrence of the string, but does require that the substituted wff be provably equivalent to the replaced string.

In what follows we make extensive use of SPE for a particular set of provable equivalents:

(a) $\neg\neg A \dashv\vdash A$ (dn).
(b) $(A \rightarrow B) \dashv\vdash (\neg A \lor B)$ (mi)
(c) $\neg(A \land B) \dashv\vdash (\neg A \lor \neg B)$ (dem)
(d) $\neg(A \lor B) \dashv\vdash (\neg A \land \neg B)$ (dem)
(e) $(A \lor (B \land C)) \dashv\vdash ((A \lor B) \land (A \lor C))$ (dist)
(f) $((A \land B) \land C) \dashv\vdash (A \land (B \land C))$ (ass)
(g) $((A \lor B) \lor C) \dashv\vdash (A \lor (B \lor C))$ (ass)
(h) $(A \land B) \dashv\vdash (B \land A)$ (com)
(i) $(A \lor B) \dashv\vdash (B \lor A)$ (com)

The rule SPE restricted to this set of provable equivalents is sufficient to enable us to transform every propositional wff to a provably equivalent wff having two useful properties:

(a)   Both its theoremhood or non-theoremhood and its validity or invalidity can be ascertained by inspection

(b)   If it is a theorem, we know how to prove it.

A.2 Conjunctive Normal Forms

Both $\vee$ and $\wedge$ are binary connectives. All conjunctions (disjunctions) are conjunctions (disjunctions) of two conjuncts (disjuncts). A conjunction of three wffs must be represented as a conjunction having a conjunction as one of its conjuncts, as, for example

$$A \wedge (B \wedge C) \text{ or } (A \wedge B) \wedge C.$$

The first conjunction is said to be *associated to the right*, the second *associated to the left*. This being said, however, for particular purposes, we can introduce a restricted convention according to which, when parentheses are omitted in either a conjunction or a disjunction of wffs, the resulting string is to be understood as associated to the left or right. We are then able to speak of a conjunction or disjunction as one of n wffs, meaning by this that when parentheses are omitted, we count n wffs separated by $\wedge$'s or n wffs separated by $\vee$'s. In what follows we will understand *disjunction (conjunction) of n wffs* to mean a disjunction (conjunction) associated to the right, which when internal parentheses are omitted, consists of n wffs separated by $\vee$'s ($\wedge$'s).

We may also speak of a conjunction or disjunction of n-terms for $n = 1$. Thus P ($\neg P$) is a conjunction of one wff; it is also a disjunction of one wff.

A conjunctive normal form (CNF) is a conjunction of disjunctions satisfying:

1. The conjunction is associated to the right.

2. The conjuncts occur in a natural order.

3. Each conjunct is a disjunction of literals satisfying

    a. The disjuncts are literals.

    b. The disjuncts occur in a natural order.

    c. All atoms precede all negations.

Thus P, ¬P and every other literal is a CNF, since it is a one-conjunct conjunction of one-disjunct disjunctions. P ∧ ¬Q is a CNF, since it is a conjunction of one-disjunct disjunctions. Equally, P ∨ ¬Q is a CNF, since it is a one-conjunct conjunction whose only conjunct is a two-disjunct disjunction.

A.3 Reduction to CNF

Any wff can be reduced to an equivalent CNF using SPE and the provable equivalents listed above.

1. Replace every occurrence of ¬¬A by A (dn).
2. Replace every occurrence of (A → B) by (¬A ∨ B) (mi).
3. Replace every occurrence of ¬(A ∧ B) by (¬A ∨ ¬B) (dm).
4. Replace every occurrence of ¬(A ∨ B) by (¬A ∧ ¬B) (dm).
5. Replace every occurrence of (A ∨ (B ∧ C)) by ((A ∨ B) ∧ (A ∨ C)) (dist).
6. Replace every occurrence of ((A ∧ B) ∧ C) by (A ∧ (B ∧ C)) (ass).
7. Replace every occurrence of ((A ∨ B) ∨ C) by (A ∨ (B ∨ C)) (ass).
8. Replace every occurrence of (A' ∧ A) by (A ∧ A') (com).
9. Replace every occurrence of (A' ∨ A) by (A ∨ A') (com).

Consider as an example, the wff

$$A: P \rightarrow (P \vee P')$$

To reduce A to CNF, we first transform it to the →-free wff by Material Implication,

$$\neg P \vee (P \vee P')$$

then commute to give

$$(P \vee P') \vee \neg P$$

then re-associate to give

$$P \vee (P' \vee \neg P).$$

which is the CNF of A.

As a second example, consider the following reduction of

$$B: (P \vee P') \rightarrow (P' \vee P)$$

$(P \vee P') \rightarrow (P' \vee P)$	
$\neg(P \vee P') \vee (P' \vee P)$	(mi)
$(\neg P \wedge \neg P') \vee (P' \vee P)$	(dm)
$(\neg P \vee (P' \vee P)) \wedge (\neg P' \vee (P' \vee P))$	(dist)
$(\neg P \vee (P \vee P')) \wedge (\neg P' \vee (P \vee P'))$	(com)
$((P \vee P') \vee \neg P) \wedge ((P \vee P') \vee \neg P')$	(com)
$(P \vee (P' \vee \neg P)) \wedge (P \vee (P' \vee \neg P'))$	(ass)

A.4 Valid CNFs: The Argument

Evidently, a CNF, A is valid iff every conjunct of A is valid. But every conjunct of A is a disjunction of literals. A disjunction of literals is valid iff for some atom B, both B and ¬B are disjuncts, for in any other case we can specify an assignment which will make the disjunction false. Therefore a CNF, A is valid iff for every conjunct of A, there is an atom, B, such that both B and ¬B are disjuncts of that conjunct. It follows that we have a mechanical test for the validity of any wff: reduce the wff to CNF and then inspect the conjuncts of that CNF. The wff is valid iff every conjunct disjoins some atom with its negation.

But then every valid CNF has a proof, since the proof of each of its conjuncts is just the proof of some substitution-instance of $P \vee \neg P$ followed by whatever applications of $\vee$-introduction, commutation and association are required, followed by applications of $\wedge$-introduction as required.

Now every wff is valid iff its CNF is valid. We can check this by checking all of the following pairs of entailments hold (They match the provable equivalences of page 243.)

(a) $\neg\neg A \dashv\vDash A$ (dn).
(b) $(A \rightarrow B) \dashv\vDash (\neg A \vee B)$ (mi).

(c) $\neg(A \wedge B)$ ⊣⊨ $(\neg A \vee \neg B)$ (dem).

(d) $\neg(A \vee B)$ ⊣⊨ $(\neg A \wedge \neg B)$ (dem).

(e) $(A \vee (B \wedge C))$ ⊣⊨ $((A \vee B) \wedge (A \vee C))$ (dist).

(f) $((A \wedge B) \wedge C)$ ⊣⊨ $(A \wedge (B \wedge C))$ (ass).

(g) $((A \vee B) \vee C)$ ⊣⊨ $(A \vee (B \vee C))$ (ass).

(h) $(A \wedge B)$ ⊣⊨ $(B \wedge A)$ (com).

(i) $(A \vee B)$ ⊣⊨ $(B \vee A)$ (com).

Since every wff is provable from its CNF, we may infer that every valid wff has a proof.

A.5 The Completeness of L: An Alternative Demonstration

The demonstration that we have just seen depends upon the equivalence of every wff of Φ to a closely specified wff expressible within a restriction of k, the set of constants of the language of L. A CNF is a very particular form which excludes all connectives other that $\neg$, $\wedge$ and $\vee$. Like the Kalmar demonstration, the latest demonstration provides a recipe for constructing a proof for any valid wff. In this section, we present a completeness demonstration for L that is more concise than the Kalmar proof or the proof using CNFs, but also less practically informative, for the method that we illustrate here merely demonstrates that every valid wff is a theorem. It provides no recipe for a proof. For many metatheoretical purposes this is sufficient since the matter of actually constructing a proof (particularly a canonical proof or a proof involving a CNF) is of less interest than the assurance that there is a proof of one sort or another.

A.5.1 An alternative strategy

We will demonstrate the weak completeness of L with respect to its truth-functional interpretation, by demonstrating that every L-consistent wff has at least one 1 in the principal column of its truth-table. Before we begin, let us remind ourselves about L-consistency and reassure ourselves that such a demonstration will actually demonstrate weak completeness.

First,

A wff is L-consistent iff ¬A is not a theorem of L.

Equivalently

A wff is L-consistent iff no contradiction is L-provable from A.

It should be evident that these two formulations are equivalent. (If we assume that ¬A is an L-theorem, then from A, we can prove A ∧ ¬A. If we assume that from A we can prove a contradiction, then by RAA we can prove ¬A from the empty set of assumptions.)
Second, let us consider the proposed strategy in detail. Suppose that we can demonstrate the following conditional thesis:

(1) For every wff A, if A is L-consistent, then the main column of the truth table of A has a 1.

Therefore, for every wff A,

(2) If ¬A is L-consistent, then the main column of the truth table of ¬A has a 1.

Then by transposition, we obtain that for every wff A,

(3) If there is no 1 in the main column of the truth table of ¬A, then ¬A is not L-consistent.

But

(4) ¬A is not L-consistent iff ⊢ ¬¬A, and so, (by DN), iff ⊢ A.

Moreover,

(5) There is no 1 in the main column of the truth-table of ¬A iff there are *only* 1's in the truth table of A, that is, iff A is a tautology.

So, for every wff A,

(6) If A is a tautology, then A is a theorem.

Accordingly, we can demonstrate that L is weakly complete by demonstrating that

(7) every L-consistent wff has a 1 in the principal column of its truth table.

> Note: The unrestrictedness of the *every* in the demonstration is not essential. With a trivial modification we can demonstrate that if every consistent wff of length less than k has a 1 in the main column of its truth table, then *every* tautology of length less than k is a theorem. (By *length* of A, we mean as usual the number of occurrences of symbols in A.)

A.5.2 The Demonstration

This demonstration, like the last, requires SPE, and exploits the fact that for every wff A of Φ, there is a wff containing no occurrences of connectives other that $\rightarrow$ and $\neg$. This follows from SPE together with sequents that we have already proved: namely

$$A \wedge B \dashv\vdash \neg(A \rightarrow \neg B)$$

and

$$A \vee B \dashv\vdash \neg A \rightarrow B.$$

The proof depends also upon the corresponding entailments, which are easily seen to hold. Thus for any wff containing occurrences of $\wedge$ or $\vee$, we can obtain a provably equivalent and semantically equivalent wff in which no $\wedge$ or $\vee$ occurs. Accordingly we need demonstrate only that every consistent wff containing no connectives other than $\neg$ and $\rightarrow$ has at least one 1 in the principal column of its truth-table. This we now proceed to do.

For the sake of brevity, we will refer to the property of having at least one 1 in the principal column of its truth table, simply as THE

PROPERTY. So we are demonstrating that every L-consistent ∧- and ∨-free wff has THE PROPERTY. The proof is a mathematical induction on the length (number of symbols) of A.

The Basis

Assume that A is of length 1. Then A is an atom (say P). But every atom of L is L-consistent, therefore A is L-consistent. Moreover, the truth table for A is

$$P$$
$$1$$
$$0$$

Thus the conditional holds for every wff of length 1.

Inductive Step

We demonstrate that
If every consistent wff A of length less than k, has THE PROPERTY, then every consistent wff A of length k, has THE PROPERTY.

Hypothesis of Induction (H.I.)

Assume that every consistent wff A of length less than k, has THE PROPERTY. Notice that this assumption is equivalent to the assumption that for every wff A of length less than k, if A is a tautology, then A is an L-theorem.

Assume that A is of length k. Then A is of the form ¬B or B → C. We consider these two cases in turn. In each we prove the transposed form of the conditional, assuming that the wff lacks THE PROPERTY, and demonstrating that it is inconsistent.

Case 1: A is of the form ¬B.

Assume that ¬B *lacks* THE PROPERTY. That is, there is no 1 in the principal column of ¬B. Then there is no 0 in the principal column of B.

Therefore, B is a tautology. Therefore, by H.I., and the earlier demonstration (restricted), ⊢ B. Therefore ⊢ ¬¬B. Therefore ¬B is inconsistent.

Case 2: A is of the form B → C.

Assume that B → C *lacks* THE PROPERTY. That is, there is no 1 in the principal column of B → C. Then there is no 0 in the principal column of B and there is no 1 in the principal column of C. Therefore A is a tautology and ¬C is a tautology. By H.I., and the earlier demonstration (restricted), ⊢ B. and ⊢ ¬C. Therefore ⊢ B ∧ ¬C. Therefore ⊢ ¬(B → C). Therefore B → C is inconsistent.

This completes the induction. We conclude that every consistent wff having no occurrences of connectives other than ¬ and → has THE PROPERTY. Therefore every wff equivalent to a wff expressible in that fragment of the language has THE PROPERTY.

We conclude that every consistent wff has THE PROPERTY, and that therefore every tautology is an L-theorem, and that therefore the system L is weakly complete.

The demonstration that L is strongly complete is the same as the demonstration following the Kalmar weak completeness demonstration.

A.5.3 Why not soundness?

Why should we not prove weak soundness of L by the converse method: by demonstrating that every satisfiable wff is L-consistent? There is no reason. This is how the demonstration will proceed:

We observe that every wff A of Φ is semantically equivalent to a wff having occurrences of no connectives other than ¬ and →. (To establish this, we need only check truth-tables.)

$$A \wedge B \ \dashv\vDash \ \neg(A \to \neg B)$$

and

$$A \vee B \mathbin{\dashv\vdash} \neg A \to B.$$

We demonstrate that every satisfiable wff A having occurrences of no connectives other than $\neg$ and $\to$ is L-consistent. The proof is by induction on the length of A.

The Basis

Assume that A is of length 1. Then A is an atom (say P). But every atom of L is satisfiable, since the truth table for A is

$$
\begin{array}{c}
P \\
1 \\
0
\end{array}
$$

But every atom is L-consistent. Thus the conditional holds for every wff of length 1.

Inductive Step

We demonstrate that if every satisfiable wff A of length less than k is L-consistent, then every satisfiable wff A of length k is L-consistent.

Hypothesis of Induction (H.I.)

Assume that every satisfiable wff A of length less than k, is L-consistent. Notice that this assumption is equivalent to the assumption that for every wff A of length less than k, if A is an L-theorem, then A is a tautology.

Assume that A is of length k. Then A is of the form $\neg B$ or $B \to C$. We consider these two cases in turn. In each we prove the transposed form of the conditional, assuming that the wff is L-inconsistent, and demonstrating that it is unsatisfiable.

Case 1: A is of the form $\neg B$.

Assume that $\neg B$ is L-inconsistent. Then B is a theorem. That is, there is no 1 in the principal column of $\neg B$. Therefore, by H.I., and the earlier demonstration (restricted), $\vDash$ B. Therefore there is no 0 in the principal column of B. Therefore $\vDash \neg\neg$B. Therefore $\neg$B is unsatisfiable.

Case 2: A is of the form B $\rightarrow$ C.

Assume that B $\rightarrow$ C is L-inconsistent. Then A is an L-theorem and $\neg$C is an L-theorem. By H.I., and the earlier demonstration (restricted), $\vDash$ B. and $\vDash \neg$C. Then there is no 0 in the principal column of B and there is no 1 in the principal column of C. Therefore there is no 1 in the principal column of B $\rightarrow$ C. Therefore $\vdash \neg$(B $\rightarrow$ C). Therefore B $\rightarrow$ C is unsatisfiable.

This completes the induction. We conclude that every satisfiable wff having no occurrences of connectives other than $\neg$ and $\rightarrow$ is L-consistent. Therefore every satisfiable wff equivalent to a wff expressible in that fragment of the language is L-consistent.

We conclude that every satisfiable wff is L-consistent, and therefore that every L-theorem is a tautology, and therefore that the system L is weakly sound.

Exercise A.42

Construct an inductive proof of weak soundness of L that does not rely upon semantic equivalences of wffs containing occurrences of $\wedge$ or $\vee$ with wffs in the $\neg$, $\rightarrow$ fragment of Φ. In this proof, four cases will be checked in the inductive step.

Exercise A.43

Construct an inductive proof of weak completeness of L that does not rely upon SPE. In this proof, four cases will be checked in the inductive step.

A P P E N D I X

B

The Connectives of Natural Language

For an introduction to connectives as symbols of a formal language, none surpasses Alonzo Church's for care and clarity. (Church 1953, 31)

> When the expressions, especially the sentences, of a language are analysed into the single symbols of which they consist, symbols which may be regarded as indivisible in the sense that no division of them into parts has relevance to their meaning, . . . there are two sorts of symbols which may in particular appear, namely primitive proper names and variables. These we may call *proper symbols.* . . But in addition there must also occur symbols which are *improper*—or in traditional (Scholastic and pre-Scholastic) terminology, *syncategorematic*—i.e., which have no meaning in isolation but which combine with proper symbols (one or more) to form longer expressions that do have meaning in isolation. . . Connectives are combinations of improper symbols which may be used together with one or more constants to form or produce a new constant.

This appendix concerns itself with connectives of formal propositional systems and more particularly the counterparts of such connectives that occur in English. In the language of a formal system, a *connective* is a symbol which, when affixed to a single sentence or to multiple sentences of the formal language, yields a sequence of symbols that constitutes a sentence of the language. In a natural language such as English the connectives form a grammatically scattered set of items that includes adverbs (*not*), coordinators and their auxiliaries (*either . . . or . . ., both . . . and . . .*, and subordinators *if . . ., only if . . .* and so on). Sensitive philosophers have found it a remarkable thing that a natural language should have such items as connectives, and though few have attempted to say in any detail how they come into human speech, it has long been recognized that logical vocabulary does not belong to that primary language in the terms of which we speak about the objects of the world. Russell:

> We pass from the primary to the secondary language by adding what I call *logical words*, such as *or*, *not*, *some*, and *all*, together with the words *true* and *false* as applied to sentences of the object language. (Russell 1940, 20)

As Russell notes, it is dauntingly difficult to say in any detail how they come to be there. Nevertheless, every known naturally occurring human language seems to have produced words that can be regarded as *logical* in character. It is also challenging to give an account of their meaning either informally or in a formal semantic theory. It is the purpose of this article to try to account for the difficulties of explaining the presence of logical vocabulary in human languages, to demonstrate how the problems are to be overcome, and to set out some of the implications, both of the difficulties and of their solutions for the practice of philosophical logic and of philosophy more generally.

B.1 Lexical Origins

All of the connective vocabulary in fact, all of the *functional* vocabulary of every natural language, has evolved from *lexical* origins. Outside the realm of logical vocabulary there are obvious examples of this: the *have* of *I have been sick*, and the *going* of *I am going to be sick* are also present, for example, in imperfect (bā) and future tense (bi) verb endings of Latin. Consider again the auxiliary *to* of the English infinitive *to be* or the auxiliary *will* of the English simple future tense. It is evident that only in very particular circumstances would such a use of *going* give rise to an expectation of locomotion for the purpose of being sick. Such uses of *going* as *I must be going crazy* are functionalized uses that have evolved from and are now distinct from an ancestral locomotive use. That connective vocabulary generally must also have had some such origin is plausibly concluded from the barest assumptions of linguistic evolution. Our species has non-linguistic ancestors. If the transition of our ancestors from a pre-linguistic condition to a linguistic condition was gradual, then we must ask   what succession of conditions likely intervened between these stages. It is evident that the elements of language with the longest genealogy must be those having ancestors capable of serving primitive functions independently of linguistic syntactic structure (whatever preconditions of behavioral or gestural syntax might also have been required). The progression from primitive to less primitive stages, we must assume, was one in which, by long successions of modest improvisations, originally relatively crude linguistic elements acquired less syntax-independent applications.

In the shorter term, we have no need for speculation or conjecture as to the origins of connective vocabulary. Without exception the connectives of current speech have descended from physical relational vocabulary of ancestral speech. No doubt the connectives of later speech will evolve from physical (often spatial and temporal) relational vocabulary of current speech. And in current speech, we find apparently transitional vocabulary, *follows from, since, in addition, as well as* and so on, that, if it does not die without issue, may eventually yield vocabulary as exclusively functional as *if, or, and* and *not*. The difficulty with the current stock of connectives is not to say from what dictionary-definable vocabulary it has descended, (that much is well understood) but rather to say how the particular character of that descent has bequeathed to our connectives the properties that they have. To illustrate the general principles that seem to govern the transition we shall have to consider some connectives (such as *but, in case, for, unless*) that are not in the usual list. However, before we embark upon that explanation, it will be as well to say something of the stock of propositional connectives that current texts take as the essential core, namely *if, not, or,* and *and*, and to give some account of the received doctrine about their uses and their meanings.

B.2 The Formal Account

Formal languages introduce their own connective vocabulary. But because historically the study of logic grew out of a study of natural language arguments, including those in which *if, not, or,* and *and* played important roles, natural languages usually provide both the motivation and the readings for the connectives introduced. Thus *not* can provide a reading for ¬; *if . . . then* for →; *or* for ∨, and *and* for ∧. Formal accounts of propositional connectives are usually provided in three parts. The first part, in the definition of Φ, the set of wffs, gives a precise and exhaustive account of the strings of symbols in which instances of the connectives can occur. The second and third parts must be understood independently of one another, and must not be confused on pain of trivializing the account. The first of these latter parts is syntactic in character, the second semantic. The first describes (in a set of rules) the roles that can be played by strings of symbols in which the connectives figure in constructions of proofs. These roles depend exclusively upon structural features of the strings, features available to visual (or auditory) inspection, without reference to interpretation of the strings or any of the elements that compose them. In the case of ∧ and ∨, these take the form of introduction and elimination rules in the system L. In the case of →, they are given by CP and MPP. In the case of ¬, they are given by the two portions of DN. It is essential that the language of *truth*, *validity* and *meaning* play no official role in our understanding of either the connectives or their rules at this stage, lest the connection with the second part of the study be made trivial. It is this second part that gives an account of the meanings of the connectives, and of the strings in which they occur; it also provides an interpretation of the rules that govern the transactions (proof constructions) into which such strings enter. Each of these three parts must be regarded as making its own contribution to our understanding of the connectives. As we shall see, each corresponds, though imperfectly, to an ingredient of our understanding of natural language connectives.

B.2.1 The syntactic preliminaries

The first element of a formal account is given by an inductive definition of the set of sentences (or *well-formed formulae* or *wffs*) of the formal language, the set Φ in the system L. The definition gives as a basis, a clause asserting that some set of *atoms* is a subset of the set of wffs, then clauses that specify how *molecular* sentences can be constructed from other sentences using the admitted connectives. So, for example, for the connectives cited, there would be clauses akin to the following.

[¬] If α is a wff, then $\neg\alpha$ is a wff.

[→] If α is a wff, and β is a wff, then $(\alpha \rightarrow \beta)$ is a wff.

[∨] If α is a wff, and β is a wff, then $(\alpha \vee \beta)$ is a wff.

[∧] If α is a wff, and β is a wff, then $(\alpha \wedge \beta)$ is a wff.

(Notice that the formation of wffs in →, ∨, and ∧ requires that parentheses be affixed.) There is generally a final (*extremal*) clause which asserts that there are no wffs other than those strings of symbols that are wffs in virtue of one or more of the previous clauses. Conventions permit the omission of outermost parentheses.

B.2.2 The proof-theoretic account

Textbook authors sometimes remark that the formation rules such as those just given serve to distinguish between strings of symbols that are meaningful and those that are not. Such remarks are no doubt intended to draw upon the reader's understanding of natural languages. But it is important to bear in mind that at that stage in the introduction of a formal language, no serious use can be made of the language of meaning. The most that such a remark can be taken to impart is that the formation rules distinguish grammatically acceptable strings from grammatically unacceptable ones. To be sure there is a kind of understanding of natural language constructions that is purely grammatical, but we can understand from internal cues how to parse a sentence without knowing what it means. We have a generally reliable idea of what nouns look like, and what verbs and so on, and some notion of their likely relative positions in a sentence given the manner in which the sentence was spoken or punctuated. Such understanding does not imply an understanding of its meaning or even that it is meaningful. If we have read Gray's Elegy, we can make a good stab at the grammar of

The curlew bowls the gal of carting hay

with no clue as to its meaning. Similar strictures apply to *conversational use* of vocabulary in natural language exchanges. That is, we can learn when to introduce it into a conversation, and learn how conversationally to react to its introduction without knowing, in any studied way, what it means. As an example, consider much of the natural language vocabulary with which philosophers have traditionally been preoccupied: *mind, idea, reality, truth, good, right, ought, deserve.* We can be perfectly fluent in our use of such vocabulary, which is to say that the vocabulary can serve all of the conversational functions to which it has become adapted without any of us having any studied understanding of it, perhaps without there being any such understanding to be had. We will return to this theme a little later; for the moment, suffice it to say that this is a reasonable view to take also of the rules for the use of the formal language connectives in proof constructions. It is common to refer to such rules as *rules of inference* but such a characterization, if taken seriously, suggests a semantic understanding of the connectives that is unnecessary for their application. It also unnecessarily introduces more philosophically puzzling vocabulary (*inference*) into the discussion. It gives us at best an illusion of additional understanding unless we can give an independent account of what inference is. As we have remarked, two such rules are usually supplied for each of the connectives of the language: a rule (called an *Introduction rule*) which permits a line to be inscribed containing occurrences of a connective having no previous occurrences, and a rule (called an *Elimination rule*) which permits a line to be inscribed containing no occurrences of a connective given some previous occurrences of the connective in lines of a proof. Thus, for example, the rule *Modus Ponens*, which permits an entry β given entries α and $\alpha \rightarrow \beta$, is an elimination rule for $\rightarrow$. The rule of *Conditional Proof*, which permits the entry of $\alpha \rightarrow \beta$ given a subproof of β from α is an introduction rule for $\rightarrow$. It would be better to understand the role of such rules by analogy with the role of instructions for the conversational use of natural language particles: those that tell us when a particle can be independently introduced, on the one hand, and those on the other hand that tell us how to respond when someone else introduces it. Of course it is no part of this account that any such rules can be usefully formulable. Consider only our understanding of the conversational role of the curious English word *but*. Of this more later.

B.2.3 The truth-theoretic account

In the formal setting, meanings of connectives are given by a *semantic theory* which specifies the conditions under which the wffs of the formal language in which the connectives occur receive one or the other of two values (0 and 1) referred to as *truth values*. Again there is room for caution. *Truth* is a term of natural language useful enough conversationally where a conversational understanding of its use is sufficient. However, it should not be assumed that *truth* and *falsity* provide more than convenient readings for the values that figure in formal semantic theories, which explicitly require of its two *truth values* only that they be distinct. Indeed the justification for the reading of these values as *truth* and *falsity* must in part depend upon the intuitive readings of the connectives and the conditions under which wffs in which they occur take one or the other of the two values. Truth-values are also exploited for the provision of standards of adequacy for arguments, again with our ordinary understanding of the goodness of arguments as its supposed basis. An argument is taken to be a good one if any conditions that make its premises true also make its conclusion true. In standard semantic theory, the atoms of a propositional language are interpreted as *propositional variables* ranging over the (truth-) values in the set $\mathbf{2} = \{0, 1\}$. Intuitively, this amounts to treating them as sentences capable of being true or false. n-ary connectives are interpreted as n-place functions that take n-tuples of truth-values as inputs and have outputs in the set $\{0, 1\}$. The input/output graphs of these functions can be tabulated in matrices such as those that follow.

$$
\begin{array}{c}
\neg A \\
0 \ 1 \\
1 \ 0 \\
\uparrow
\end{array}
$$

$$
\begin{array}{ccc}
A \rightarrow B & A \wedge B & A \vee B \\
1 \ \ 1 \ 1 & 1 \ 1 \ 1 & 1 \ 1 \ 1 \\
1 \ \ 0 \ 0 & 1 \ 0 \ 0 & 1 \ 1 \ 0 \\
0 \ \ 1 \ 1 & 0 \ 0 \ 1 & 0 \ 1 \ 1 \\
0 \ \ 1 \ 0 & 0 \ 0 \ 0 & 0 \ 0 \ 0 \\
\uparrow & \uparrow & \uparrow
\end{array}
$$

The matrix for $\rightarrow$ requires some explanation. On the intuitive reading for the $\rightarrow$, it might be wondered whether conditions for its truth and falsity could adequately be registered in such a table. The justification for the interpretation lies in metatheoretic demonstrations that all and only arguments that are good on the understanding outlined above can be given proofs in the system. But it can also be made plausible by an appeal to the intuitive readings of the connectives. An inspection of the matrix for $\vee$ reveals that a disjunction $\alpha \vee \beta$ is true if and only if at least one of its *disjuncts* α, β is true. We can express this as the conditional 'If one of the disjuncts is not true, then the other disjunct is'. Thus the truth-condition for the disjunction $\neg\alpha \vee \beta$ can be given as 'if $\neg\alpha$ is not true, then β is true'. But on our intuitive understanding of 'not', which is reflected in the matrix for $\neg$, $\neg\alpha$ is not true if and only if α is true. Accordingly the truth-condition for $\neg\alpha \vee \beta$ can be given as 'If α is true then β is true', and the matrix for $\alpha \rightarrow \beta$ is exactly that of $\neg\alpha \vee \beta$. The formal truth-theory of the connectives is said to be *compositional*, by which may be understood that it interprets every wff as a particular composition of truth-functions. For example, it interprets the wff $\neg$ p $\vee$ (q $\rightarrow$ r) as a truth-function, taking as its input the ordered triple of the values of p, q, r, and giving as its output a value in $\{0, 1\}$. Its output value can be computed by giving to $\vee$ the pair of values consisting of the output of $\neg$ at p on the one hand and the output of $\rightarrow$ at the ordered pair <q, r> on the other.

B.2.4 Consequences of the formal account

The connectives of a formal theory are in certain respects well behaved. In the first place, the rules for the formation of well-formed formulae, since they require parentheses, guarantee that every wff of the language is grammatically unambiguous. As a consequence, sentences of a formal language are compositional both with respect to applications of rules and with respect to the computation of truth-values. By the former is meant that, for example the rule of $\vee$-introduction permits the inscription of $\alpha \vee \beta$ as a line of a proof given that α is already a line, without regard for the complexity of α or the complexity of β. This is what makes a formal language *formal*, for it enables us to speak of the *form* of a wff α, and to identify that form with the set of wffs that can be composed by uniformly substituting wffs for occurrences of atoms in α. In the semantic theory, compositionality guarantees that meanings are stable under all such substitutions. The stability also enables us to speak confidently of conditionals, negations, disjunctions, and conjunctions. A conditional is any wff of the form p → q; a negation is any wff of the form ¬p and so on. Moreover, it matters little, except pedagogically, when we speak of disjunctions, for example, whether we speak of the syntactic notion (a wff of the form p ∨ q) or the semantic conception (a wff that has the value 1 exactly when at least one of its components has the value 1). The metatheory proves that the two notions coincide.

B.3 The Connectives of Natural Language

By the standards of formal languages, the connectives of natural language are systemically ill-behaved. In the first place, sentences of natural languages are not always syntactically unambiguous, and the ambiguity may or may not make a significant difference to the conditions under which we would regard a sentence as true. Contrast the sentence[9]

No trees have fallen over here

on the reading

Over here no trees have fallen

with the same sentence on the reading

Here no trees have fallen over.

With some effort we might find a difference in the truth-conditions, but in practical terms, it does not matter which sentence is intended, and we might use the sentence without noticing the ambiguity or without caring much about it if we did. Sometimes syntactic ambiguity does make a difference. Consider[10]

Most of the voters voted for most of the candidates

9. The example is due to Mary Shaw.

10. The example is due to Peter Geach.

which is ambiguous as between claiming of most of the candidates that most of the voters voted for them, and claiming of most of the voters that they voted for most of the candidates. If there are five voters and five candidates, then conditions can be such that three out of five of the voters vote for at least three candidates, but that only two candidates receive more than two votes. Secondly, natural languages are not compositional in ways that would parallel the compositionality of formal languages. To be sure, the significance of any sentence of a natural language is in some rough manner a function of its parts and the way in which they have been combined, and in that measure natural languages are compositional, but the nature of the compositionality is not such in general as to give rise to anything corresponding to the uniform substitutional notion of form available in a formal language. The sentence

She is at home or she is at the office

is plausibly labelled a disjunction, both because it can be expected to support roles in inferences that closely correspond to those that a wff of the form p ∨ q could play in the construction of a proof, and because it will correctly be supposed true if an only it is supposed that at least one of its component sentences is true. But the sentence

She may be at home or she may be at the office

permits the inference that she may be at home, and permits the inference that she may be at the office. It therefore supports an inferential role that is more akin to the role of ∧ in proofs than to that of ∨. For this reason, the uniform substitutional definition of *form* does not apply to sentences of natural language, and the question as to whether a given sentence is, for example, a disjunction or a conjunction must be understood semantically rather than syntactically. When philosophers speak of 'the logical form' of a natural language sentence, it is safe to assume that they have in mind some correct formalization of the sentence, given its presumed truth-conditions.

B.3.1 Natural language connectives with dual meanings

As understood in classical propositional logic, conjunction and disjunction are *duals*. Proof-theoretically this means that $(\neg\alpha \wedge \neg\beta)$ and

¬(α ∨ β) are interderivable as are (¬α ∨ ¬β) and ¬(α ∧ β). Semantically, a matrix equivalent to the truth-matrix for ∨ can be obtained from the truth-matrix for ∧ and vice versa by a thorough-going interchange of 1's and 0's. Thus to say that English *or* sometimes behaves like English *and* can be expressed, albeit crudely, by saying that *or* has dual meanings or dual uses. In fact to express the matter this way is a simplification of the case. In the presence of modal sentence elements, *or* behaves like *and*, but the kind of conjunction is of a rather specialized sort that mediaeval logicians labeled *ut nunc*, meaning roughly *as things are now*. If I say, 'You may have tea or you may have coffee', I give you permission to have tea, and I give you permission to have coffee, but I do not commit myself to permission to have both; we can express that restriction by saying that by this formula, I give you permission, as things are now, to have tea, and I give you permission, as things are now, to have coffee. However, if things change, as for example, if you accept coffee, the permission to have tea may lapse, and if things change through your accepting tea, the permission to have coffee may lapse. It is only *as things are now* that both permissions are in effect.

It should not be assumed that this *ut nunc* character is entirely absent from sentences that might seem candidates for representation in propositional logical notation, for a similar stricture applies to certain conditional constructions. For example if I say 'If Laurie volunteers or Alison volunteers, I'll be very much surprised' this too may be an *ut nunc* claim. Such a sentence is, as we would expect, equivalent to 'If Laurie volunteers, I'll be surprised, and if Alison volunteers, I'll be surprised', so if you believe me, you might reasonably predict that Laurie's volunteering will surprise me, and again you might reasonably predict that Alison's volunteering will surprise me, but I might make such a claim even if I knew that Alison's volunteering invariably prompts Laurie to volunteer and vice versa. Once again, since assurances evaporate once the conditions are relevantly altered, the conjunction of conditionals to which we take that conditional to be equivalent must be regarded as *ut nunc*. If Laurie volunteers, then Alison's volunteering may not occasion surprise.

Notice too that in such cases, we can, and sometimes do, express even the conjunction of conditionals using an *or* instead of an *and*, as 'If Laurie volunteers, I'll be surprised, or if Alison volunteers, I'll be surprised (too)'. On comparatively rare occasions we hear *or* used conjunctively when the sentences that form the conjuncts are negated, as Sam Donaldson's complaint (about some presidential candidates) 'They don't come on *Face the Nation* or they don't appear on *Meet the Press*'. That *or* should have such conjunctive uses, but only in certain linguistic environments is a significant datum that an account of natural language connectives must somehow explain. In fact this is a feature of numerous natural language connectives.

Virtually no natural language connectives have only a single use, but certain of them have pairs of uses that bear to one another a relationship akin to the relationship between conjunction and disjunction. We can list a few other examples by way of making the general point, whilst indicating some theoretically significant variations. Consider first the word *but*. We are told by some textbook authors to represent sentences composed with *but* as conjunctions, and in many cases, such as 'Mary arrived on time, but Hugh was early', that is the best we can do. However, some sentences composed with *but* must be represented as disjunctions, as for example, 'It doesn't rain but it pours', which on its historical reading means 'If it rains, it pours', that is, 'Either it doesn't rain or it pours.' A second less obvious example is the dual uses of *for* which sometimes has the effect of *because of* ('For your crime you must be punished') and sometimes that of *in spite of* ('For all my efforts I didn't succeed'). To see the connection between the two readings, compare the two approximately equivalent parsings of 'We shan't get to the bar for all the kids': first, 'We shan't (get to the bar for (*i.e., in spite of*) all the kids)', and second, '(We shan't get to the bar) for (*i.e., because of*) all the kids'.

The phenomenon is not restricted to the connectives that we associate with those of propositional logic. Generally speaking we represent the quantitative adjective *any* by the use of a universal quantifier. But *any*, like *or* in its conjunctive uses, appears only in certain environments (We say 'I haven't read any of those books', but we do not say 'I have read any of those books'; We say 'She can ask any student', but we do not say 'She asks any student'.) and it sometimes demands representation as an existential quantifier. Now that we *can* sometimes represent sentences having *any* by wffs containing existential quantifiers is not controversial, precisely because of the duality of (x) and (∃x). We can represent 'I haven't read any of those books' either by (x)(Fx → ¬ Gx) or by ¬(∃x)(Fx ∧ Gx).

The fact that in certain cases a conditional sentence having an occurrence of *any* in its *if*-clause can be represented equivalently with long-scope (x) and with short-scope (∃x) is a common topic of introductory texts, which point out that the quantificational wff (x)(Fx → P) is provably equivalent to the wff (∃x)Fx → P. However, there are English conditional sentences in which, in spite of this equivalence, distinguish existential *any* from universal *any*. Consider the sentence 'If any student touches my whisky, I'll know it' in which the reference of the *it* is ambiguous. On one reading the sentence claims that I will know the identity of the poacher; on the other reading it claims only that I will know of the poaching. On either reading, it could receive representation with long-scope (x), but the two readings would be distinguished along the lines of the distinction between (x)(Fx → K_a(Fx)) and (x)(Fx → K_a((∃y)Fy)). That there is a reading of the sentence that demands the second representation is demonstration proof that *any* sometimes requires representation by an existential quantifier. The problem besets other vocabulary that, while it is not on our short list of logical words, nevertheless occurs in sentences that can be represented in the more usual logical vocabulary. *Without* is an example. Contrast the following pair of sentences in the first of which the *without* is naturally represented as *if . . . not*, and in the second of which as *and . . . not*: 'She'll die without medical attention'/'She'll die without disclosing the secret'. Neither is this *and/if* phenomenon restricted to modern English. Any adequate account of natural language connectives must explain why, in Jacobean English *and* had acquired uses like those of *if*, as in 'Nay, and you go conjuring I'll be gone' (Marlowe *Faustus* **x**, 76).

B.3.2 Dualization

The explanation for these dualities (the term is intended here to cover all cases resulting from similar historical developments) can be illustrated by reference to the familiar English connective *unless*. Like other connectives, *unless* is a collapsed residue of a longer construction, in this case, the construction *on [a condition] less than that* (the bracketted portion included only to give the sense.) Now that construction of fourteenth- and fifteenth-century English would be represented as, roughly, *and not*. α on a condition less than that of β, is α and not β. Yet that use is now lost, replaced by uses which we are told by textbooks to represent by *or*, or *if . . . not*. How did this happen? The answer seems to be this: In its earliest uses, the longer ancestral construction is not known to have occurred outside of environments that were negative in character. Now just as two people might agree in their use of the earlier-quoted sentence 'No trees have fallen over here', that is, use it on the same occasions and to the same effect, yet disagree in their syntactic construal of it, so two generations of language users can agree in their uses of the ancestral *unless* construction, but disagree in their syntactic construal of these constructions. The results of such a syntactic schism can be illustrated schematically. Consider the string 'Not α unless β' where α and β represent clauses. Now consider the construals 'Not (α unless β)' and '(Not α) unless β' where the parentheses indicate the relative scopes of *not* and *unless*. Suppose that one linguistic sect takes the former as their scope arrangements with *unless* being understood as *and not*, and the other, while agreeing with the former on the correct occasions of use of such constructions, nevertheless construes the scope arrangements according to the second scheme. If this second sect must learn from these uses how *unless* itself is to be construed, then it must learn the disjunctive construal, and if, under its stewardship, the use of *unless* migrates to environments that are not negative in character, in those environments it will *require* the disjunctive reading, a reading which that sect must be proportionally sufficient in the population of language users to enforce. But that reading now accords with more instances of *unless* than the older reading, and so becomes a plausible standard for later generations of language users. Now that is a greatly simplified account of a kind of mutation that takes place in the evolution of connective vocabulary. It certainly accounts for some of the systematic diversity of meaning that particular connectives exhibit. For example, it is certainly the explanation for the appearance in the language of the earliest conjunctive uses of *but*. Nor, more generally speaking, is it a particularly rare *kind* of linguistic development. At the level of word formation it has given us 'an umpire' where we had 'a numpire', 'a nickname' where we had 'an ick name' and so on. Canadian parliamentary practice has produced a charming instance. Before any provincial or national election is called, the premier or prime minister must draw up a writ, and this activity has traditionally been reported metonymously as the intention to ask for a dissolution of parliament and the fixing of an election date. However, the *draw up* of this construction has been heard as *drop*, and in this connection, the cognate forms of *drop* have replaced those of *draw up*. So now one hears tell of the PM being about to *drop* the writ or as *dropping* the writ or as having just *dropped* the writ. The arcane nature of the activities involved in the calling of an election shelter the obscurity of the construction, which is no doubt associated for some with a document's being dropped off at the Governor General's house, and for others with the dropping of a flag at the beginning of a competition. Most likely the survival of the idiom is indirectly due to the familiarity of *dropping the puck*, the action which starts a hockey match. The actual history of connective use is perhaps complicated beyond our capacity to give detailed accounts even of their present use, and they all exhibit much greater and more subtle diversity in their natural habitat than their formal counterparts can be expected to convey. Merely to explain how mutations of the kind that we have outlined here almost certainly does not explain even all of the instances of duality we have listed let alone the many other natural language uses. Conjunctive *or* is one case in point. It seems likely that, whatever role syntactic mutation may have played, other developments have greatly complicated the story. There is, however, one other case in which mutation has played a part in producing diversity of connective meaning. Once the phenomenon has brought itself to our attention in other cases, it becomes apparent that English is in possession of two *if*'s, one of which is a mutation of the other.

B.3.3 Two *if*'s of English

Because formal languages are primarily written languages, negation in formal languages presents no noticeable difficulty. A sentence, however complex in its structure is negated by the prefixing (or suffixing) to it of a unary connective. In natural languages there is generally no such prefixable negator. Philosophers, with great state, pronounce *It is not the case that...* (and some, by extension, can be heard to say *It is the case that...*, but in the working world of speech, no such elaborate construction would likely survive in day-to-day use. In general English negates complex sentences by placing a *not* somewhere in the vicinity of the main verb, usually after an auxiliary *do*, as 'He goes/He does not go', and lets the context or order or plausibility constraints cue its role as that of negator of the main clause or negator of the whole sentence. (Consider 'I'm not standing here now because of my political convictions'/'Because of my political convictions, I'm not standing here now'.) In the case of the conditional, fewer conventions have been established to distinguish the two. This is not to say that there are no unambiguous ways of expressing negations of conditionals, and we have recourse to such construction-types as 'Just because α, that doesn't mean that β'. And even with a single syntactic form, we can generally force distinctions *prosodically*, that is, by a variation of pitch contour, by stress and by syllable lengthenings. Thus we can distinguish between the claim that winning a lottery will make Fred miserable and the claim that even winning the lottery will not raise Fred from his misery by saying in the former case, 'If he wins the lottery, he'll be *miserable*' and in the latter 'If he wins the *lottery* he'll be miserable'. We can reinforce the latter with an initial *even*, but we need not do so. Now those sentences represent distinguishable uses of English *if*: the former a *sufficiency* use, and the latter an *insufficiency* use. The evidence suggests that this bifurcation has come about in part because of the negation placement in conditionals such as 'If you were the Queen of England, I couldn't let you into this facility' where the *if*-clause represents an extreme illustrative case from which conclusions about lesser cases are to be drawn. We can compare such a case with that of 'You realize, don't you, that if you were the Queen of England, I couldn't let you into this facility, but because you are the night janitor I must' in which the same clause is merely contrasted with the present case. The first conditional asserts the insufficiency of a case (your being the Queen of England) for admission to the facility, and it does it by negating a sufficiency *if*-sentence. The second has a reading on which it has a negated main clause, and asserts the sufficiency of being the Queen of England for exclusion from the facility. What initiates the development of an insufficiency *if* is the assimilation of the syntax of the former to the syntax of the latter that preserves the distinction between their correct occasions of use. Here as elsewhere, the rule is *New Syntax plus Old Satisfaction-conditions implies New Connective Meaning*. Schematically, the construction, which would justify a representation as $\neg(\alpha \rightarrow \beta)$ retains the same class of occasions of use, but is read as if having the structure $\alpha (\rightarrow) \neg\beta$. The reading of $(\rightarrow)$ that justifies that syntactic understanding is for a time concealed beneath the reading of $\rightarrow$ that matches the older syntactic construal. But when it is the standard syntactic reading for a sufficient proportion of a sufficiently large population of language users, then the *if* with its newer, insufficiency meaning can migrate to negation-free environments with only its prosodic presentation, or possibly that reinforced by the addition of *even* to distinguish the use. In English, then there are at least those two uses of *if*: one which I have labelled the sufficiency *if*, which could perhaps also be called the *implies if*, and the one that I have labelled the insufficiency *if*, but which could also be labelled the *does not imply that not if*. Expressing the distinction in this implicational idiom serves to point up a remarkable feature of the insufficiency *if*, namely that it is *hybridized*; in particular, although the insufficiency *if* represents a substantially new use, nevertheless, it has also inherited formal characteristics from the sufficiency *if* that are at odds with its new reading. The point is best made by comparing its behavior with that of what ought to be its formal representation. Observe first that

$$(\alpha \vee \beta) \rightarrow \gamma$$

is equivalent to the conjunction

$$(\alpha \rightarrow \gamma) \wedge (\beta \rightarrow \gamma).$$

Accordingly the negated

$$\neg((\alpha \vee \beta) \to \neg\, \gamma)$$

is equivalent to

$$\neg((\alpha \to \gamma) \wedge (\beta \to \neg\, \gamma))$$

and hence, by De Morgan's theorems, to

$$\neg(\alpha \to \gamma) \vee \neg\, (\beta \to \neg\, \gamma).$$

Yet insufficiency *if*'s of English behave just as sufficiency *if*'s behave with respect to disjunctive antecedents. We might argue unexceptionably:

> If the Queen (herself) asked me I wouldn't disturb Mrs. Smith;
> if the King (himself) asked me I wouldn't disturb Mrs. Smith;
> so if the King asked me *or* the Queen asked me, I wouldn't disturb
> Mrs. Smith.

The presence of the negations seems to trigger the use of *or*, without detailed compositional justification: if α is insufficient, and β is insufficient, then $\alpha \vee \beta$ is insufficient; that a negated conditional serves as a vehicle for the expression of the insufficiency plays no compositional role. But notice that in certain respects, the behavior of the insufficiency *if* is consonant with the behavior of a negated conditional. The Queen's asking might be insufficient, and the King's asking likewise insufficient, but it does not follow that a joint request would not tip the case. So we cannot correctly argue:

If the Queen (herself) asked me I wouldn't disturb Mrs. Smith; if the King (himself) asked me I wouldn't disturb Mrs. Smith; so if the King asked me *and* the Queen asked me, I wouldn't disturb Mrs. Smith

even though such an argument would in general be unexceptionable for the sufficiency *if*. No doubt we can construct formal semantic theories for a hybrid conditional such as the one we have here briefly noted. But the question arises here as elsewhere in natural language whether the construction of a formal semantic theory brings with it the kind of understanding appropriate to the subject matter. The alternative would be an explanatory theory of natural language, one which offers *inter alia* a general account of how connectives evolve and of the forces that shape their evolution. The struggle to understand is in part a struggle against oversimplifications. It is simply an error to suppose that because we are perfectly fluent in our use of the connectives (or, come to that, any other device of natural language) that we have any deeper understanding of it than is required for that fluency itself. And because our linguistic ancestors were in a similar condition, it is (perhaps less simply) an error to suppose that we can have any but a very flawed understanding of the present diversity of uses or of the historical processes that produced it. But it is a corollary of these observations that we should not permit ourselves to be seduced by the elegance or even the usefulness of a formalism into the notion that our ordinary understanding of it gives us authority to speak of natural language.

B.4 Disjunction

B.4.1 Introduction

(**Note**: This section repeats, for the specific case of disjunction, sufficient of the material of the previous sections that it can be read independently.) A disjunction is a kind of compound sentence historically associated by English-speaking logicians and their students with indicative sentences compounded with *either . . . or*, such as

Either I am very rich or someone is playing a cruel joke.

But nowadays the term *disjunction* is more often used in reference to sentences (or well-formed formulae) of associated form occurring in formal languages. Logicians distinguish between the abstracted *form* of such sentences and the roles that sentences of that form play in arguments and proofs — the *meanings* that must be assigned to such sentences to account for those roles. The former represents their *syntactic* and *proof-theoretic* interests and their *semantic* or *truth-theoretic* interest in disjunction. Introductory logic texts are sometimes a little unclear as to which should provide the defining characteristics of disjunction. Nor are they clear as to whether disjunctions are primarily features of natural or of formal languages. Here we consider formal languages first.

B.4.2 Syntax

The definition of a formal system, either axiomatic or natural deductive, requires the definition of a language, and here the formal vocabulary of disjunction makes its first appearance. If the disjunctive constant $\vee$ (historically suggestive of Latin *vel (or)*) is a primitive constant of the language, there will be a clause, here labelled [$\vee$] in the inductive definition of the set of well-formed formulae (wffs). Using α and β as metalogical variables, ranging over wffs, such a clause would read:

[$\vee$] If α is a wff and β is a wff, then $\alpha \vee \beta$ is a wff

perhaps accompanied by an instruction that $\alpha \vee \beta$ is to be referred to as the *disjunction* of the wffs alpha and beta, and read as "<name of first wff> vel (or 'vee', or 'or') <name of second wff>". Thus, on this instruction, the wff 'p $\vee$ q' is the *disjunction* of p and q, and is pronounced as 'pea vel queue' or 'pea vee queue' or 'pea or queue'. In this case, 'p' and 'q' are the disjuncts of the disjunction.

If $\vee$ is a non-primitive constant of the language, then typically it will be introduced by an abbreviative definition. In presentations of classical systems in which the conditional constant $\rightarrow$ or (mi) and the negational constant $\neg$ are taken as primitive, the disjunctive constant $\vee$ might be introduced in the abbreviation of a wff $\neg\alpha \rightarrow \beta$ ($\neg\alpha$ (mi) β) as $\alpha \vee \beta$. Alternatively, if the conjunctive $\wedge$ has already been introduced either as a primitive or as a defined constant, $\vee$ might be introduced in the abbreviation of a wff $\neg(\neg\alpha \wedge \neg\beta)$ as $\alpha \vee \beta$.

B.4.3 Proof theory

Much as we would understand the conversational significance of vocabulary more generally if we had a complete set of instructions for initiating its use in a conversation, and for suitable responses to its introduction by an interlocutor, we give the proof-theoretic significance of a connective by providing rules for its introduction into a proof and for its elimination. In the case of $\vee$, these might be the following:

[$\vee$-introduction]

For any wffs α and β, a proof having a subproof of α from an ensemble Σ of wffs, can be extended to a proof of $\alpha \vee \beta$ from Σ.

[$\vee$-elimination]

For any wffs α, β, γ, a proof that includes (a) a subproof of $\alpha \vee \beta$ from an ensemble of wffs Σ, (b) a subproof of γ from an ensemble $\Delta \cup \{\alpha\}$, and a subproof of γ from an ensemble $\Theta \cup \{\beta\}$ can be extended to a proof of γ from $\Sigma \cup \Delta \cup \Theta$. Intuitively, the former would correspond to a rule of conversation that permitted us to assert A or B (for any B) given the assertion that A. Thus if we are told that Nicholas is in Paris, we can infer that Nicholas is either in Paris or in Toulouse.

Intuitively, the latter rule would correspond to a rule that, given the assertion that A or B, would permit the assertion of anything that is permitted both by the assertion of A and by the assertion of B. For example, given the assertion on certain grounds that Nicholas is in Paris or Toulouse, we are warranted in asserting on the same grounds plus some geographical information, that Nicholas is in France, since that assertion is warranted (a) by the assertion that Nicholas is in Paris together with some of the geographical information and (b) by the assertion that Nicholas is in Toulouse together with the rest of the geographical information. More generally we may sum the matter up by saying that the rule corresponds to the conversational rule that lets us extract information from an *or*-sentence without the information of either of its clauses. In the example, we are given the information that Nicholas is in Paris or Toulouse, but we are given neither the information that Nicholas is in Paris nor the information that he is in Toulouse.

B.4.4 Semantics

In its simplest, classical, semantic analysis, a disjunction is understood by reference to the conditions under which it is true, and under which it is false. Central to the definition is a *valuation*, a function that assigns to every atomic, or unanalysable sentence of the language a value in the set {1,0}. In general the inductive truth-definition for a language corresponds, clause by clause, to the definition of its well-formed formulae. Thus for a propositional language it will take as its basis a clause according to which an atom is true or false accordingly as the valuation maps it to 1 or to 0. In systems in which $\vee$ is a primitive constant, the clause corresponding to disjunction takes $\alpha \vee \beta$ to be true if at least one of α, β is true, and takes it to be false else. Where $\vee$ is introduced by either of the definitions earlier mentioned, that truth-condition can be computed for $\alpha \vee \beta$ from those of the conditional ($\rightarrow$ or (mi)) or conjunction ($\wedge$) and negation ($\neg$).

Now the truth-definition can be regarded as an extension of the valuation from the atoms of the language to the entire set of wffs with 1 understood as the truth-value *true*, and 0 understood as the truth-value *false*. Thus, classically, disjunction is semantically interpreted as a binary truth-function from the set of pairs of truth-values to the set {0, 1}. The tabulated graph of this function, as dictated by the truth-definition, is called the truth-table for disjunction. That table is the following:

$$
\begin{array}{ccc}
A & \vee & B \\
1 & 1 & 1 \\
1 & 1 & 0 \\
0 & 1 & 1 \\
0 & 0 & 0 \\
  & \uparrow &
\end{array}
$$

B.4.5 Inclusive and exclusive disjunctions

Authors of introductory logic texts generally take this opportunity to distinguish the disjunction we have been discussing from another binary truth-function $\underline{\vee}$ whose graph is tabulated by the table:

$$A \veebar B$$
$$1\ 0\ 1$$
$$1\ 1\ 0$$
$$0\ 1\ 1$$
$$0\ 0\ 0$$
$$\uparrow$$

where $\alpha \veebar \beta$ is read α xor β. This truth-function is referred to variously as exclusive disjunction, as 0110 disjunction (after the succession of values in its main column), and as logical difference. The wff $\alpha \veebar \beta$ is true when exactly one of α, β is true; false else. To make matters explicit, the earlier discussed truth-function $\vee$ is called *inclusive*, or *non-exclusive* or *1110* disjunction.

B.4.6 Natural language

It is an assumption, at any rate a claim, of many textbook authors that there are both uses of *or* in English that correspond to 1110 disjunction and uses that correspond to 0110 disjunction, and this supposition generally motivates the introduction and discussion of the xor connective. Since we are following the usual order of textbook exposition, this is perhaps the moment to make a few observations on this score. The first are purely syntactic. The *or* of English that such authors cite is a coordinator (or coordinating conjunction). It can coordinate syntactic elements of virtually any grammatical type, not merely whole sentences. Moreover, if we consider only its uses joining whole sentences, we must notice that it can join sentences of virtually any mood: interrogative sentences and imperatives as well as indicative sentences can be joined by *or* in English. And again, if we restrict our attention to its uses joining indicative sentences, we must note that *or* is by no means restricted to the binary cases in this role. Indeed, there is no theoretical finite limit to the number of clauses that it can join. This is perhaps the most fundamental relevant syntactic difference between *or* on the one hand and $\vee$ and $\veebar$ on the other. The sentence

Nathalie has been and gone or Nathalie will arrive today
or Nathalie will not arrive at all

is a perfectly correct sentence and not ambiguous as between

(Nathalie has been and gone or Nathalie will arrive
today) or Nathalie will not arrive at all

and

Nathalie has been and gone or (Nathalie will arrive today
or Nathalie will not arrive at all).

By contrast, the wff $p \vee q \vee r$, far from being ambiguous as between $(p \vee q) \vee r$ and $p \vee (q \vee r)$, is, on the inductive definition of well-formedness, not a wff. If the parenthesis-free notation is tolerated in general logical exposition, this is because $\vee$ is *associative*, that is, the wffs $(p \vee q) \vee r$ and $p \vee (q \vee r)$ are syntactically interderivable, and semantically have identical truth-conditions. The formal account of disjunction could readily be liberalized to accommodate that fact, and even conveniently in languages in which $\vee$ was primitive. In that case our inductive definition of the language could permit any such string as $\vee(\alpha_1, \ldots, \alpha_i, \ldots, \alpha_n)$ to be well-formed if $\alpha_1, \ldots, \alpha_i, \ldots$ and α_n are. The relevant clause of the truth-definition would accordingly be modified in such a way as to give $\vee(\alpha_1, \ldots, \alpha_i, \ldots, \alpha_n)$ the maximum of the truth-values of $\alpha_1, \ldots, \alpha_i, \ldots$ and α_n. Moreover, this accords well with such cases as the one cited in which *or* joins more than two simple clauses: such a sentence is true if at least one of its clauses is true; false otherwise.

The fact that English *or* is not binary does not accord so well with the claim made by many textbook authors that there are uses of *or* that require representation by 0110 disjunction. To be sure, $\underline{\vee}$ is associative, so that a notational liberalization would be possible, parallel to the one described for $\vee$. Syntactically the claim could be accommodated. But, as Hans Reichenbach (1947) seems first to have pointed out, the truth-definition for $\underline{\vee}(\alpha_1, \ldots, \alpha_i, \ldots, \alpha_n)$ would have to be such as to give it the value 1 if any odd number of $\alpha_1, \ldots, \alpha_i, \ldots, \alpha_n$ have the value 1; the value 0 otherwise. The result is evident from the truth-table where $n > 2$. For $n = 3$, suppose that $(\alpha \underline{\vee} \beta) \underline{\vee} \gamma$ has the value 1. The truth-definition as given by the table requires that exactly one of $\alpha \veebar \beta$, γ has the value 1. Let γ have the value 1; then $\alpha \veebar \beta$ has the value 0. Then α and β have the same value. That is, either both α and β have the value 0, or both α and β have the value 1. In the former case exactly one of α, β, γ has the value 1; in the latter, all three have the value 1. That is, the disjunction will take the value 1 if and only if an odd number of disjuncts have the value 1. A simple induction will prove that this result holds for an exclusive disjunction of any finite length. It is sufficient for present purposes to note that, in the case where n = 3, $\veebar(\alpha_1, \alpha_2, \alpha_3)$ will be true if all of its disjuncts are true. Now there is no naturally occurring coordinator in any natural language matching the truth-conditional profile of such a connective. There is certainly no use of *or* in English in accordance with which five sentences A, B, C, D, and E can be joined to form a sentence A or B or C or D or E, which is true if and only if either exactly one of the component sentences is true, or exactly three of them are true or exactly five of them are true.

Most of the texts make no claims about exclusive disjunctive uses of either English or Latin *or*-words beyond the two-disjunct case. But it is a fair presumption that the belief in exclusive disjunctive uses of *or* in English includes just such three-disjunct uses of *or*. Such a use of *or*, would be one in accordance with which three sentences A, B, and C can be joined to form a sentence A or B or C, which is true if and only if exactly one of the component sentences is true. Though not a 0110-disjunctive use of *or*, this would be a general use representable as 0110 disjunction in the two-disjunct case.

The question as to whether there is such a use of *or* in English, or any other natural language goes to the very heart of the conception of truth conditional semantics. For it seems certain that there are conversational uses of *or* that invite the inference of exclusivity, but which do not seem to require exclusivity for their truth. Thus, for example, if one says (Tarski 21) 'We are going on a hike or we are going to a theater', even with charged emphasis upon the *or*, one will have spoken falsely if in the event we do both, unless, as in Tarski's example, one has also denied the conjunction.

Some authors have sought examples of 0110 disjunction in *or*-sentences whose clauses are mutually exclusive, as that of Kegley and Kegley (232).

John is at the play, or he is studying in the library

of which they remark, 'There is no mistaking the sense of *or* here: John cannot be in both places at once'. If their example were an example of exclusive disjunction, we could safely infer from it that the play is not being performed in the library, that the theatre is not in the library, that John is not swotting in the stalls between acts while his companion fights her way to the bar to fetch the drinks. In fact, even, perhaps particularly, when the disjuncts are genuinely mutually exclusive, there are no grounds for the supposition that the *or* represents 0110 disjunction. Were there such grounds the $\vee$ of formal logic would require distinct semantic accounts for the wffs $p \vee q$ and $p \vee \neg p$. As Barrett and Stenner point out (p 3), the case requires quite the reverse. Since the truth-tables of $\vee$ and $\underline{\vee}$ differ exactly in the output value of the first row, what alone would clinch the case for the existence of an exclusive *or* would be a sentence in which both disjuncts were true, and the disjunction therefore false. No author has yet produced such an example.

B.5 The myth of *vel* and *aut*

If the claims of the generality of logic texts dictate the structure and content of our discussion, it is perhaps as well to dispel another current myth—namely that the notational choice of $\vee$, (read as *vel*) as the connective of inclusive disjunction, and the claim that the English *or* has 0110-disjunctive uses are supported by the facts of the Latin language. Copi (241) is as explicit as any:

The Latin word "vel" expresses weak or inclusive disjunction, and the Latin word "aut" corresponds to the word "or" in its strong or exclusive sense.

The idea is, first, that whereas English has only one *or*-word, Latin has two: *vel* and *aut*, and secondly, that the uses of *vel* in Latin would be representable as 1110 disjunction and the uses of *aut* as 0110 disjunction. As to the first, the very shape of the claim is likely to mislead. The case is not that Latin had two words for *or*, but rather that Latin had more than one word that gets translated into English as *or*. In fact, Latin had *many* words that are translated into English as *or*, including, besides the two listed, at least *seu, sive* and the enclitic *ve*. So does English have many words that can be translated into English as *or*, including *unless, if . . . not, but* (It does not rain but it pours) and so on. All vocabulary has a history, and languages accumulate vocabulary that becomes adapted to nuanced uses.

Now the supposition that Latin had a 0110 coordinator must suffer from the same implausibilities as the corresponding supposition about English. What of the two-disjunct case? If any general tendency can be detected in actual Latin usage, say in the classical period, that would distinguish the uses of *vel* from those of *aut*, it is that *aut* tended to be brought into use in the formation of lists of disjoint or contrasted or opposed items, categories or classes or states, as for example

Omne enuntiatum aut verum aut falsum est
(Every statement is either true or false) (Cicero *De Fato*).

The difficulty with these examples is that the exclusiveness of the states independently of the choice of connective must mask any disjointness that the connective could itself impose. That it does not impose *any* disjointness itself is best seen in its list-forming uses. Consider the list (Cicero *De Officiis*) *tribunos aut plebes*, (the magistrates or the mob, *accusative plural*). To be sure the categories are disjoint, and this fact might be supposed to contribute to the selection of *aut*. But the mutual exclusion in such cases need not survive the addition of a verb. *Timebat tribunos aut plebes* (one feared the magistrates or the mob) does not exclude the case in which one feared both. However, what clinches the refutation of this mythical supposition is that if that whole clause is brought within the scope of a negator, the resulting sentence will expect a reading along the lines of negated 1110 disjunction. *Nemo timebat tribunos aut plebes* (No one feared the magistrates or the mob) just means no one feared either. It does not mean everyone either feared neither or feared both (in other words, that they feared the one if and only if they feared the other). Since the negation of a 0110 disjunction is a 1110 disjunction (either both disjuncts are true or both disjuncts are false), this use of *aut* cannot be a 0110 disjunctive use.

In fact, in classical Latin, *aut* was favoured over *vel* in constructions involving negations, and in that use, *aut* behaves analogously to ∨. But pretty well anywhere an *aut* could be used, a *vel* could be substituted, and vice versa. The resulting sentence would have a different flavour, and in some instances would be mildly eccentric, but would not have a different truth condition. The uses of *vel* reflected its origins as an imperative form of *volo*. The flavour of *Nemo timebat vel tribunos vel plebes* would be closer to that of *Name which group (of the two) you will: no one feared them. Aut* was adversative: no one feared either social extremity. (For more examples and a more detailed discussion, see Jennings *The Genealogy of Disjunction*, 239-251.)

B.5.1 The *or* of natural language

There are undoubtedly disjunctive uses of *or* in English, and of corresponding vocabulary in other natural languages. But the uses of *or* after the pattern of the logic texts: (*Either Argentina will boycott the conference or the value of lead will diminish* and so on) constitute only a very small proportion, certainly fewer than 5% of the occurrences of *or* in English, and, it can be supposed, of corresponding words in all other natural languages as well. It is therefore not surprising that it should be some of these non-disjunctive uses that have been misidentified as instances of exclusive disjunction. The example cited by T.J. Richards, (84) is a good representative example of one such common misidentification:

> So how can we find a clear-cut case of the exclusive 'or'? Imagine a boy who asks for icecream *and* strawberries for tea. He is told as a sort of refusal: 'You can have icecream *or* strawberries for tea'. Here there is no doubt: not both may be had.

Once again there is a difficulty in trying to account for the exclusivity by reference to truth-conditions, though, if we are permitted to consult the intentions of the speaker (as Richards himself does) we may be in no doubt as to the prohibition of strawberries and icecream, however curious such a prohibition might seem. But this example, in company with the many others like it (which this author has sometimes referred to collectively as *the argument from confection*) suffers from the even more serious flaw that it is not a disjunction at all. The problem is not that the *or* does not join whole clauses. Even if we expand the example to 'You can have icecream for tea *or* you can have strawberries for tea', the sentence cannot be construed as a disjunction. The reason is that the child would be correct in inferring that he can have icecream for tea, and would be correct in inferring that he can have strawberries for tea. Such sentences are elliptical for conjunctions, not for disjunctions, even on a truth-conditional construal. It just happens that for such conjunctions, questions of exclusivity, or rather non-combinativity also arise.

Not every *or* of English (nor every counterpart of *or* in other languages) is disjunctive, even among those that join pairs of indicative sentences.

References
Robert B. Barrett and Alfred J. Stenner,
    "The Myth of the Exclusive 'Or'", *Mind* **80** 116-121.

Church, Alonzo,
    Introduction to Mathematical Logic. Princeton: Princeton University Press, 1956.

Cicero, Marcus Tullius,
    De Fato, Translated by H. Rackham. Cambridge, Mass: Harvard University Press, 1942.

Cicero, Marcus Tullius,
    De Officiis, Translated by Walter Miller. Cambridge, Mass: Harvard University Press, 1975.

Copi, I.M.,
    Introduction to Logic, New York: Macmillan, 1971.

Geach, Peter,
    Reference and Generality, Ithaca, New York: Cornell University Press, 1980.

Jennings, R.E.,
   The Genealogy of Disjunction, New York: Oxford University Press, 1994.

Horn, Laurence,
   The Natural History of Negation. Chicago: University of Chicago Press, 1989.

Kegley, Charles W. and Jacquelyn Ann Kegley,
   Introduction to Logic, Lanham, MD: University Press of America, 1984.

Marlowe, Christopher
   The Tragical History of Dr. Faustus, 1604.

Reichenbach, Hans.,
   Elements of Symbolic Logic, New York: Macmillan, 1947.

Richards, T.J.,
   The Language of Reason, Rushcutters Bay, NSW, 1978.

Russell, Bertrand,
   An Inquiry into Meaning and Truth, London: Allen and Unwin, 1940.

Alfred Tarski,
   Introduction to Logic and to the Methodology of Deductive Sciences, New York: Oxford University Press, 1941 (Revised 1946 edition).

Chapter One

Logicianer

(p. 6) A student of logic.

Chapter Two

Argument

(p. 8) An ensemble of sentences, one of which is designated (usually by an *illative* modifier) as conclusion, and the rest are premisses.

Illative

(p. 8) Pertaining to inference. Illative modifiers are sentence modifiers that indicate that one sentence is being inferred. Examples: *therefore, so, accordingly, It follows that . . ., then.* Some subordinators, such as *since, because* are also referred to as illatives, because of their use in reports of inferences, but they might better be called *premissives* because they indicate premissings rather than concludings.

Formal system

(p. 12) (Preliminary account) A formal natural deductive system S is a pair comprising: (a) a language and (b) a set of rules for extending proofs.

Metalogical variable

(p. 14) A variable in the *metalanguage q.v.* of a language and ranging over objects of some uniform type.

Metalanguage

(p. 14) The language in which a language or a formal system which it underlies is described, discussed, or theorized about.

S-proof

(p. 15) (Preliminary account) An S-proof of a sentence A from an ensemble of sentences Σ is a finite sequence of sentences in the language of S (the entries of the proof), each of which is justified by a rule of S, and the last of which, A, requires only assumptions occurring in Σ.

S-sequent

(p. 15) Any finite sequence of sentences $A_1, \ldots A_n$ in the language of S, followed by an S-turnstile ($\vdash_S$), followed by a single sentence, A_{n+1} in the language of S. An S-sequent can be understood as a claim that there is an *S-proof* of the sentence following the turnstile from the ensemble of sentences that precede it.

S-provability

(p. 16) (Preliminary account) A sentence, A is S-provable from an ensemble Σ if and only if there is an S-proof of A from Σ. S-provability is a relation consisting of the set of pairs Σ, A for which A is provable from Σ.

Assumption

(p. 16) An entry of a proof that is justified by the Rule of Assumption.

Sequent corresponding to a line of a proof

(p. 17) The sequent corresponding to line (i) of a proof is the sequent having the assumptions of line (i) to the left of its ⊢ (as its premises) and the entry of line (i) to the right of its ⊢ (as its conclusion).

Reflexivity

(p. 18) A property of S-provability (for a formal system S) that A is S-provable from an ensemble Σ of sentences, if A is a member of Σ. S-provability is said to be *reflexive*. The system L has this property in virtue of the Rule of Assumption, which permits any sentence to be L-proved from itself, and *monotonicity q.v.*

Structural property

(p. 18) A property of S-provability, the statement of which makes no mention of particular connectives in the language of S.

Structural rule

(p. 18) A rule of a formal system S, which makes no mention of particular connectives in the language of S.

Negation

(p. 20) A unary operation on wffs, the output of which for a wff A is ¬A. The wff resulting from such an operation.

Discharge of an assumption

(p. 21) Invariably, a proof that is an input to a rule begins with an assumption. A rule that takes a proof (or proofs) as its input is a rule that permits the extension of a proof from a proof of a conclusion that requires an assumption to a proof of a conclusion that does not. When a proof is so extended by such a rule, the assumption is said to have been discharged.

Introduction rule

(p. 24) A rule that permits the extension of a proof to a proof of a sentence having occurrences of a connective of which there are no earlier occurrences in the proof.

Elimination rule

(p. 24) A rule that permits the extension of a proof to a proof of a sentence having no occurrences of a connective of which there are earlier occurrences in the proof.

Monotonicity

(p. 26) A property of S-provability (for a formal system S) that if A is S-provable from an ensemble Σ, then A is S-provable from any ensemble that includes Σ. S-provability is said to be *monotonic*.

Contradiction

(p. 32) The conjunction of any sentence A with its negation, in that order: $A \wedge \neg A$.

Transitivity

> (p. 34) A property of S-provability (for a formal system S) that if A is S-provable from an ensemble Σ, then any sentence B that is S-provable from Σ, A is S-provable from Σ. S-provability is said to be *transitive*.

Compactness

> (p. 34) (also called *finiteness*) A property of S-provability (for a formal system S) that if A is S-provable from an ensemble Σ, then A is S-provable from some finite ensemble that is included in Σ. S-provability is said to be *compact* or *finite*.

Finiteness
> (p. 34) See *compactness*

Interderivability

> (p. 39) Two sentences A and B are interderivable (in a system S) (A ⊣⊢ B) if and only if there is an S-proof of B from A, and an S-proof of A from B. Any such pair of sentences are also *deductive equivalents q.v.*

Deductive equivalents

> (p. 39) Two sentences A and B are deductive equivalents (or deductively equivalent) if and only if anything provable from A is provable from B and *vice versa*.

Chapter Three

Formal system

> (p. 49) (Final account) A formal natural deductive system S is a pair <*L, R*> of which *L* is a *language q.v.*, and *R* a set of rules for extending a proof.

Metatheorem

(p. 49) A sentence of the metalanguage of a formal system which asserts some demonstrable property of the formal system.

Language

(p. 50) The language underlying a formal system is a triple $<At, k, \Phi>$ of which At is a set of atoms, k a set of constants, and Φ a set of well-formed formulae (wffs).

Inductive definition

(p. 51) A definition of a set which identifies some subset (the basis of the definition) and gives an exhaustive set of operations under which the set is *closed q.v.*.

Closed set

(p. 51) A set X is closed under an n-ary operation $*$ if and only if $x_1, \ldots, x_n$ are members of X, then $*(x_1, \ldots, x_n)$ is a member of X.

Closure

(p. 51) The closure of a set X under a set of operations $*_1, \ldots, *_n$ is the smallest set X′ that includes X and is closed under the operations $*_1, \ldots, *_n$.

Conditionalization

(p. 51) A binary operation on wffs. The conditionalization of a wff B on the wff A has the wff $A \to B$ as its output. The wff resulting from such an operation.

Conjunction

> (p. 51) A binary operation on wffs. The conjunction of wff A and wff B has the wff A ∧ B as its output. The wff resulting from such an operation.

Disjunction

> (p. 51) A binary operation on wffs. The disjunction of wff A and wff B has the wff A ∨ B as its output. The wff resulting from such an operation.

Scope

> (p. 52) The scope of an occurrence of a connective is the shortest wff of which that occurrence is a part.

Binding power

> (p. 52) The binding power of a connective is the conventionally established comparison of the scope of its occurrences with those of occurrences of other connectives. Conventions of binding power permit the omission of brackets. A convention gives one connective greater binding power than another if and only if, by that convention there is a bracket-free wff containing occurrences of both is read in such a way that the scope of an occurrence the former is included in the scope of the latter.

S-proof

> (p. 53) (Final account) An S-proof of a wff A from an ensemble of wffs Σ is a finite sequence of wffs in the language of S (the entries of the proof), each of which is justified by a rule of S, and the last of which, A, requires only assumptions occurring in Σ.

S-theorem

(p. 53) An S-theorem (theorem of S) is a wff that is S-provable from the empty (null) ensemble of assumptions.

Uniform substitution

(p. 55) (Propositional version) An operation on wffs and sequents. A wff B (a sequent S′) is obtained by uniform substitution from a wff A (a sequent S), if and only if B (S′) is the result of uniformly substituting an occurrence of some wff C for every occurrence of some atom in A (S).

Substitution instance

(p. 56) A wff B is a substitution instance of A if and only if B is obtainable from A by some sequence of uniform substitutions.

Theorem introduction

(p. 58) A derived rule by which theorems already proved are cited as justifications of lines of proofs. The use of theorem introduction can be thought of as an abbreviative device by which we omit proofs of substitution instances of theorems already proved.

Sequent introduction

(p. 61) A derived rule by which sequents already proved are cited as justifications of lines of proofs. The use of sequent introduction can be thought of as an abbreviative device by which we omit proofs of substitution instances of sequents already proved.

Arithmetization

> (p. 69) An arithmetization is a translation of each of the sentences of a language, and perhaps also each of the proofs of a system uniquely into a numeral of arrithmetic. Since the set of such numerals has a natural order, namely the order of the associated natural numbers, the translation gives an enumeration of the sentences or proofs in which each such sentence or proof has a place finitely far along.

Validity of an argument

> (p. 71) On the most usual account of validity, an argument is said to be valid if and only if the truth of its premisses guarantees the truth of its conclusion.

Propositional variable

> (p. 72) A variable that ranges over the set of *truth-values q.v.* The standard interpretation of an atom of propositional logic is as a propositional variable.

Truth-values

> (p. 72) Numerical values in the set $\{1, 0\}$, read as *true* and *false* respectively.

Truth-value assignment

> (p. 72) An assignment of 1 or 0 to every atom.

n-ary truth-function

> (p. 73) A function that takes an n-tuple of truth-values as its input and gives a truth-value as its output.

Graph of a function

(p. 73) The graph of a function f is the set of the input/output pairs of f.

Truth-table

(p. 73) The tabulation of the graph of a truth-function. It displays all input values to the function as well as the output value for each of those inputs.

Material conditional

(p. 74) A binary truth-function that outputs the value 0 for the input $<1,0>$ and outputs a 1 for all other inputs.

Compositions of functions

(p. 75) The composition of two functions f and g is the function $g \circ f$, where $g \circ f(x)$ is $g(f(x))$, that is $g \circ f$ is the function that outputs the outputs of g for inputs that are the outputs of f. Complex wffs are interpreted as compositions of truth-functions.

Constant truth-function

(p. 76) A truth-function whose output is the same for every input.

Tautologous truth-function

(p. 77) A constant truth-function whose output is the value 1 for every input. Also called *tautological truth-function*.

Unsatisfiable truth-function

(p. 77) A constant truth-function whose output is the value 0 for every input.

Tautologous wff

(p. 77) A wff whose interpretation is a tautologous truth-function. Also called *tautological wff*.

Inconsistent wff

(p. 77) A wff from which a contradiction is provable. Also called *inconsistency*.

Entailment

(p. 83) Generally, an ensemble Γ of wffs *entails* a wff A on an interpretation if and only if if every wff of Γ is true in the interpretation, then A is true in the interpretation. Since in the present case, sentences are made made true by assignments of truth-values to atoms, the relevant notion of validity can be stated as: an ensemble Γ *entails* a wff A ($\Gamma \vDash A$) if and only if every assignment of truth-values to the atoms of the language that makes all of the wffs of Γ true, makes A true. Also called *semantic entailment*.

Valid wff

(p. 83) A wff is valid (in an interpretation) if and only if, on that interpretation it is entailed by the empty set. In propositional logic, the set of valid wffs is identical to the set of tautologies.

Strong soundness

(p. 83) A formal system S is said to be strongly sound if and only if for every ensemble Σ of wffs, and every wff A, if A is S-provable from Σ, then Σ entails A.

Weak soundness

(p. 83) A formal system S is said to be weakly sound if and only if for every theorem is a valid wff. In particular, L is weakly sound if and only if every L-theorem is a tautology.

Strong mathematical induction

(p. 84) A method of demonstrating that every member of an enumerable set has some property. Such a demonstration requires demonstrating, first, that the first member of the set has the property, and second, that if every object before the ith object has the property, then so does the ith object.

Consistent system

(p. 91) A system S is consistent if and only if no contradiction is a theorem of S.

Strong completeness

(p. 92) A formal system S is said to be strongly complete if and only if for every ensemble Σ of wffs, and every wff A, if Σ entails A, then A is S-provable from Σ.

Weak completeness

(p. 92) A formal system S is said to be weakly complete if and only if for every valid wff is a theorem. In particular, L is weakly complete if and only if every tautology is an L-theorem.

Literal

(p. 93) A literal is a propositional wff that is either an atom or the negation of an atom.

Sequent corresponding to a row of a truth-table

(p. 93) The sequent corresponding to the row of a truth-table of a wff A is the sequent having, (a) on the left of its $\vdash$, every atom of A having a 1 in that row and the negation of every atom having a 0 in that row, and (b) on the right of its $\vdash$, the wff A if A receives a 1 in that row, or the negation of A if A receives a 0 in that row. (In the special case in which A is a tautology, every row corresponding to a row of the truth-table of A has A as its conclusion.)

Lemma

(p. 94) A minor result demonstrated as a preliminary to the demonstration of a relatively major result.

Canonical sequent

(p. 99) A sequent corresponding to a row of a truth-table of a tautology. If the tautology has occurrences of n distinct atoms, then it has 2^n canonical sequents, each of which has that tautology as conclusion.

Conditional corresponding to a sequent

(p. 102) The conditional corresponding to a sequent is the conditional obtained by iterating the operation of (a) conditionalizing the conclusion of the sequent on the rightmost premiss and then (b) deleting that premiss, until no premisses remain.

Maximality

(p. 105) A property of a formal system. S is a *maximal* formal system if and only if the addition of any new wff to the set of its theorems yields an inconsistent system. The set of L-theorems is a maximal consistent uniform-substitution-closed set.

Chapter Four

n-place predicate symbol

> (p. 108) An n-place predicate symbol is an upper case letter of the Roman alphabet. It is taken to represent a property instantiated by n (not necessarily distinct) individuals. Thus, for example, the P of Px is taken to represent a one-place property, that is, a property such as *squareness* capable of instantiation by a single individual; the P of Pxy is taken to represent a two-place property, that is, a property such as *betterness* capable of instantiation only by a pair of individuals (of which one is better than the other); the P of Pxyz is taken to represent a three-place property, that is, a property such as *betweenness* capable of instantiation only by a triple of individuals (one of which is between the other two) and so on. A propositional atom is a zero-place predicate symbol.

Proper name

> (p. 108) Any of the lower case letters l, m, n and so on from the middle portion of the Roman alphabet. A proper name may be intuitively understood as the actual name of a particular individual as distinct from a name assigned to whatever object instantiates some complex of properties.

Individual variable

> (p. 109) The individual variables of a formal system (x, y, z and so on) are variables ranging over individuals of some or other domain. They may be thought of playing the role in a formal language akin to the role of pronouns in a natural language.

Universal quantifier

> (p. 109) Any individual variable enclosed in brackets, as (x), (y) and so on. The quantifier (x) is read 'For all x . . .' or 'It is true as regards every x that . . .' or some variant.

Existential quantifier

> (p. 110) Any individual variable preceded by a reverse 'E', (∃), all enclosed in brackets, as (∃x), (∃y) and so on. The quantifier (∃x) is read 'For some x . . .' or 'There exists at least one x such that . . .' or some variant.

Arbitrary name

> (p. 120) One of a, a′, a″ and so on. The proof-theoretic counterpart of an arbitrary object, that is, an object about which no assumptions are made beyond those explicitly stated. Thus, for example an arbitrary isosceles triangle is an object about which nothing is assumed (for example as to size or shape) except those required for isosceles triangularity. In proof-theory an arbitrary name is distinguished from a proper name in its use in assumptions prompted by existentially quantified sentences, and in the sufficiency for universal generalizations of some sentences having no names but arbitrary names.

Chapter Five

Term

> (p. 148) A term is any proper or arbitrary name.

Nominal

> (p. 148) A nominal is any term or individual variable.

Symbol

(p. 148) A *symbol* is either a bracket or a term or an individual variable or a predicate letter (including sentence letters) or reverse-E.

Formula

(p. 148) A *formula* is any finite sequence of symbols.

Atomic sentence (quantificational)

(p. 149) If $t_1, \ldots, t_n$ are terms (not necessarily distinct), for $n \geq 0$, and P a predicate letter, then $Pt_1 \ldots t_n$ is an *atomic sentence*.

Quantificational well-formed formula

(See p. 149 for this definition.)

Propositional function

(p. 151) A formula A is a *propositional function in the variables* $v_1, \ldots, v_n$, for $n \geq 0$, if $(v_1) \ldots (v_n)A$ is a wff.

Scope of a quantifier

(p. 153) The scope of an occurrence of a quantifier in a propositional function is the shortest propositional function in which it occurs.

Side condition

(p. 154) A side condition is a (typically non-syntactic) requirement imposed by a rule for extending a proof. Side conditions on EE and UI restrict their use to cases in which the appearance of arbitrary names in required assumptions are circumscribed or prohibited. See page 156 for the details.

Quantificational uniform substitution

> (See p. 163.)

Variable replacement

> (p. 167) A wff B results from a wff A by variable replacement if
> and only if there are occurrences of the variable v in A, but no
> occurrences of the variable v′, and B is the result of replacing
> every occurrence of v in A by an occurrence of v′.

Alphabetic variants

> (p. 166) Wffs A and B are alphabetic variants if and only if B is
> obtainable from A by iterating the operation of *variable
> replacement q.v.*.

Quantificational model

> (p. 169) A quantificational model is a structure consisting of a
> *domain of objects*, an interpretation of n-ary predicate symbols as
> n-ary relations on the objects of the domain, and terms are
> interpreted as names of objects of the domain. Unquantified wffs
> receive truth values accordingly as the objects named by their
> terms have the properties named by their predicate symbols. A
> universally or existentially quantified wff **A** (in the variable v) are
> true or false accordingly as (respectively) all or some of the objects
> of the domain have the property that interprets the longest
> propositional function in v occurring in **A**.

Quantificational validity

> (p. 171) A wff is *valid* on a domain, D if and only if it is true in
> every model on D. A wff is *universally valid* if and only if it is
> valid on every domain.

First-order Theories

(p. 176) The label *first-order* is applied to a language in which only individual variables have occurrences in quantifiers. A *first-order theory* is a theory expressible in a first-order language, hence *first-order theory of identity, first-order theory of relations* and so on.

Definite description

(p. 183) The label applied by Bertrand Russell to certain occurrences of definite noun phrases, as in *the author of Waverly*.[11] Such occurrences are conventionally taken to imply both existence and uniqueness.

Enthymeme

(p. 190) An *enthymeme* is an argument in which either the conclusion or a significant premiss is implicit.

Modal Logic

(p. 196) A logic the underlying language of which contains '$\square$' or related connectives. Generally speaking, modal logics are logics taken to represent notions of necessity, possibility, and so on.

---

11. A curious piece of literary gossip is given in the "Montreal Herald" of 15th July, 1820. It is a letter from a correspondent, stating that Thomas Scott, brother of Walter Scott. and then serving in Canada as paymaster of the 70th Regiment, was the author of "Waverly," "The Antiquary," "Guy Mannering" and "Rob Roy." The writer asserts that he saw the the manuscript of "The Antiquary" in Thomas Scott's handwriting, and that portion relating to Flora McDonald in the handwriting of Mrs. Scott. (Quoted from J.F. Pringle. *Lunenburgh or the Old Eastern District.* Cornwall, Ontario: Standard Printing House, 1890.)

A, B, C,…: 10, 14, 50

$P_1, P_2, P_3,…$: 14, 50

P, Q, R…: 14, 50

$\rightarrow$:14, 50, 52, 149

$\vee$:14, 50, 52, 149

$\wedge$:14, 50, 52, 149

$\neg$:14, 50, 52, 149

$\Gamma, \Delta, \Sigma…$:16

$\vdash_L$:16

$\varnothing$:16

A:17, 45

MPP: 18, 45

DN: 20, 45, 66

CP: 20, 45

$\wedge$I: 24, 46

$\wedge$E: 25, 46

$\vee$I: 28, 46

$\vee$E: 29, 47

C: 47, 125, 155

RAA: 32, 47

$\leftrightarrow$: 35, 51, 52

df.: 36

MTT: 40,64

MPT: 42,

DeM: 44,

$\vdash$: 53

$\dashv\vdash$: 39, 52

$\dashv$: 39

L: 49

R: 49

S: 49

At: 50

k: 50

Φ: 50

wff: 50, 149

$B_1,…,B_n$: 53

LEM: 54, 65, 64

Id: 64

TI: 58

TI(S): 58

$A_1,…,A_n$: 56

SI: 61

SI(S): 61

MI: 65

MTP: 65

DS: 65

Con: 66

τ: 70

1 (truth value): 72

0 (truth value): 72

$g \circ f$: 75

|: 82

$\downarrow$: 82

$\models$: 83, 211

H.I.: 85

$W_1,…,W_n$: 95

$CS1 - CS2^n$: 99

F, G: 108

l, m, n: 108

x, y, z: 109

(x), (y), (z): 109

$(\exists x), (\exists y), (\exists z)$: 110

INDEX *of* EXERCISES

Student number: Philosophy instructor
email suffex: capcollege.bc.ca
Email: sgardner @ capcollege.bc.ca

new Professor